T0305947

Portfolio Management

Portfolio management is becoming the 'must have' for organizations to prosper and survive in this decade and beyond. No longer can the organizational focus be one of following best and repeatable practices as resource limitations mean only those programs, projects, and operational work that add business value can and should be pursued. Executives are focusing on strategic ability and managing complexity, which can only be done through a disciplined portfolio process in ensuring the best mix of programs, projects, and operational work is under way. In turn, the portfolio is constantly in flux as difficult decisions are made if a project, for example, is no longer contributing to business value and providing benefits and should be terminated to reallocate resources to one of higher priority. Commitment to this difficult approach is necessary at all levels, and communication is required so everyone knows how their work contributes to the organization's strategic goals and objectives.

Portfolio Management: Delivering on Strategy, Second Edition focuses on the benefits of portfolio management to the organization. Its goal is to provide senior executives a view on how portfolio management can deliver organizational strategy. The emphasis is on the specific aspects within the portfolio management discipline and how each aspect should be managed from a business perspective and not necessarily from a portfolio management perspective. Highlights of the book include:

- Agile portfolio management
- Delivering organizational value
- Portfolio management and uncertainty
- Portfolio governance
- Marketing a portfolio
- Portfolio management success

Starting with a review of the project portfolio concept and its development, this book is a reference for executives and practitioners in the field, as well as a students and researchers studying portfolio management.

Best Practices in Portfolio, Program, and Project Management

Leading Virtual Project Teams: Adapting Leadership Theories and Communications Techniques to 21st Century Organizations
2nd Edition

Margaret R. Lee
April 6, 2021

Project Business Management
1st Edition

Oliver F. Lehmann
July 31, 2020

Marketing Projects
1st Edition

Olivier Mesly
February 07, 2020

Project Management Methodologies, Governance and Success: Insight from Traditional and Transformative Research
1st Edition

Robert Joslin
April 22, 2019

Implementing Project and Program Benefit Management
1st Edition

Kenn Dolan
November 26, 2018

Culturally Tuning Change Management
1st Edition

Risto Gladden
October 09, 2018

https://www.routledge.com/Best-Practices-in-Portfolio-Program-and-Project-Management/book-series/CRCBESPRAADV

Portfolio Management
Delivering on Strategy
Second Edition

Edited by
Carl Marnewick
John Wyzalek, PfMP

CRC Press
Taylor & Francis Group
Boca Raton London New York

CRC Press is an imprint of the
Taylor & Francis Group, an **informa** business
AN AUERBACH BOOK

Cover Image Credit: Shutterstock.com

Second edition published 2023
by CRC Press
6000 Broken Sound Parkway NW, Suite 300, Boca Raton, FL 33487-2742

and by CRC Press
4 Park Square, Milton Park, Abingdon, Oxon, OX14 4RN

CRC Press is an imprint of Taylor & Francis Group, LLC

© 2023 Taylor & Francis Group, LLC

First edition published by CRC Press in 2014

ISBN: 978-1-032-32629-0 (hbk)
ISBN: 978-1-032-32626-9 (pbk)
ISBN: 978-1-003-31590-2 (ebk)

DOI: 10.1201/9781003315902

Typeset in Minion
by SPi Technologies India Pvt Ltd (Straive)

Contents

Preface

Portfolio management is becoming the 'must have' for organizations to prosper and survive in this decade and beyond. No longer can the organizational focus be one of following best and repeatable practices as resource limitations mean only those programs, projects, and operational work that can add business value and should be pursued. Executives are focusing on strategic ability and managing complexity, which only can be done through a disciplined process in ensuring the best mix of programs, projects, and operational work is under way. In turn, the portfolio is constantly in flux as difficult decisions are made if a project, for example, is no longer contributing to business value and providing benefits and should be terminated to reallocate resources to one of higher priority. Commitment to this difficult approach is necessary at all levels, and communication is required so that everyone knows how their work contributes to the organization's strategic goals and objectives.

Even though the organizational strategy changes, a roadmap of the portfolio can show the work in progress, its dependencies with other work, its benefits, and constraints. An approach then to exploit and embrace change is needed along with a dedication within the organization to be the market leader even though one's own work may not be necessary. The goal is to emphasize business value, benefits, and commitment to strategy with critical and creative thinking core competencies at all levels.

Portfolio Management: Delivering on Strategy, Second Edition focuses on the benefits of portfolio management to the organization. Its goal is to provide senior executives a view about how portfolio management can deliver organizational strategy. The emphasis is on the specific aspects within the portfolio management discipline and how each aspect should be managed from a business perspective and not necessarily from a portfolio management perspective. Highlights of the book include the following:

- *Project portfolio management for product development.* A review is given on the development of new products in the context of a portfolio delivering on organization strategy.

- *Developing a new portfolio.* The necessary role of change management in creating a new portfolio is examined.
- *Portfolio governance.* Chapter 4 explores the six key questions that the organization needs to answer to develop a pragmatic, sensible, and effective GM. Chapter 5 reviews the structure and functions of portfolio management.
- *Marketing a portfolio.* How to win the support of the various stakeholders is a discussion on marketing the portfolio benefits and persuading stakeholders that the commitment they are going to make, a price they pay, is worth the benefit to them.
- *Portfolio management success.* Understanding and analyzing how the success of a portfolio is managed is important to understanding what portfolio management entails, which the chapter reviews.
- *Portfolio communications management.* One chapter on this topic reviews many aspects of communications management in a portfolio setting and explains how communication should be implemented around strategic portfolio management in order to maximize its impact and relevance throughout the organization. A second reviews the roles of the governance board, portfolio manager, and related stakeholders in portfolio communications management.
- *Managing a portfolio in an uncertain environment.* The zone of uncertainty is presented as well as potential remedies for disruption and disappointment that occurs in the zone. These remedies are principally related to improving relationships among stakeholders.
- *Agile portfolios.* Insight is given on how organizations can ensure that agile initiatives align to the organizational strategy.
- *Dynamic capabilities in portfolio management.* Portfolio management challenges are created in dynamic environments, and dynamic capabilities aid portfolio managers in identifying and managing these challenges. Examples illustrate aspects of portfolio management that help organizations respond to change and improve organizational outcomes.

Starting with a review of the project portfolio concept then covering portfolio management trends and techniques, the book is designed as a reference for executives and practitioners in the field. The book can also serve as a textbook for university courses on portfolio management.

Contributors

Amaury Aubrée-Dauchez is an experienced senior executive with a 28+ year proven career track record in delivering business-centric solutions that achieve organization strategic needs. In 2010, he created a management consulting firm, Webbed Star – now Koosai – providing strategic advisory services to prestigious clients like Nestlé, Services Industriels de Genève, Groupe Mutuel, Canton of Valais, or UBS for achieving excellence and adopting industry standards. Besides, he helps founders and investors to unrig the unicorn gameplay.

Jennifer Young Baker is an associate professor at the University of Southern California and assistant professor at Wingate University. She is a management consultant who has been managing projects, programs, portfolios, and PMOs for 30 years for Fortune 500 companies and government entities as well as authoring her own books and articles. She was featured in İpek Sahra Özgüler's book, *The Perspective of Women in Project Management.* Jennifer holds a MSPM, PMP, PgMP, PfMP, ITIL, BRMP, SSMBB, CB-PMO, and SAFe-SA certifications. She lives in North Carolina with her husband Ken and two Yorkies.

Dr. Lynda Bourne DPM, FACS is a senior management consultant, professional speaker, teacher, and an award-winning project manager focused on the delivery of stakeholder management and sharing her experience in books, academic, and conference papers over the 50 years of her career. In 2010, she was engaged as visiting professor at EAN University, Colombia, teaching in the Masters of PM Program for five years. Most recently, she's been a lecturer in IT management at Monash University, in particular focused on stakeholder engagement, communication, and leadership

Dr. Wanda Curlee has over 30 years of project/program/portfolio management experience. Her business career spans the areas of IT, Government, Insurance, Telecommunications, and Utilities. Dr. Curlee is active with the Project Management Institute (PMI). She was part of the team that developed the Portfolio Management Professional (PfMP) certification and panel review. Dr. Curlee has published three books; two delve into complexity theory and one is about virtual PMOs. She is proud to be the mother of three children who have served or currently serve in the US Military. She met her husband, Steve, while both served in the US Navy.

Julie Delisle is an Assistant Professor in Project Management at the School of Management (ESG) from University of Quebec in Montreal (UQAM). She teaches project and agile management in the graduate programs in project management. A former entrepreneur, she has worked in a wide range of fields (IT, mining, marketing, etc.), and in a wide range of roles related to project management (Agile coach, consultant, advisor, etc.) Her research focuses on tensions, paradoxes, and practices in organizations, with a special interest for temporality, project organizing and agile contexts. Her work has been published in international peer-reviewed academic journals as *International Journal of Project Management and Project Management Journal*.

Professor Catherine P. Killen [University of Technology Sydney (UTS)] has degrees in mechanical engineering and engineering management and a PhD in project portfolio management. Her academic career builds on industry experience in product development, innovation, and the introduction of new technologies. Catherine has published more than 90 papers in books, conferences, and journals such as the *International Journal of Project Management, Project Management Journal*, and *IEEE Transactions on Engineering Management*. Catherine's current research interests include strategic multi-project management practices; visualizations and portfolio decision-making; emergent strategy and portfolio-level processes; and the establishment of sense-making capabilities to improve project management practice.

Professor Carl Marnewick's academic career started in 2007 when he joined the University of Johannesburg. He traded his professional career as a senior Information Technology (IT) project manager for that of an academic career. The career change provided him the opportunity to emerge himself in the question why IT/IS-related projects are not always successful and do not provide the intended benefits that were originally anticipated. This is currently a problem internationally as valuable resources are wasted on projects and programs that do not add value to the strategic objectives of the organization. It is an international problem where there is a gap between theory and practice, and he is in the ideal position to address this problem. The focus of his research is the overarching topic and special interest of the strategic alignment of projects to the vision of the organizations.

Yvan Petit, MEng, MBA, PhD, PMP, PfMP, ACP, is a full professor at the Business School of the University of Quebec at Montreal (ESG UQAM) since 2010. After being the program director for the post-graduate programs in project management, he was the associate dean for international relations at ESG UQAM between 2018 and 2021. He has taught in Canada, France, Algeria, Vietnam, and Sweden and in 2017 he received the Teaching Innovation Award at UQAM. His research interests are portfolio management, agile approaches, and uncertainty management. He has over 25 years of experience in project management, primarily in software development and R&D in the telecommunications industry. He has served as a member of the PMI Standards MAG (Member Advisory Group) and on the Canadian section for ISO standards on Project Management.

Alejandro Romero-Torres is a full professor at the School of Management (ESG) from University of Quebec in Montreal (UQAM), co-holder of the chair in project management ESG UQAM, director of the Lab for innovative practices in project context and director of the Observatory of public projects. His main research expertise is the governance and management of infrastructure and digital projects, including issues related to stakeholders' engagement, decision-making, and value creation. He works closely with public organizations in Quebec to sustain project performance. He is also Director of the Global Center Accreditation GAC of the Project Management Institute. He has a doctorate and master's degree in technology management from Polytechnique de Montréal and a bachelor's degree in information technology engineering from Anahuac University (Mexico).

David Tennant, MBA, MS, PE, PMP is a sought after business consultant, speaker, and trainer. He is an engineer by training and a manager by experience with a natural gift for relating to people. His clear, direct, yet light-hearted communication style engenders trust across business functions as well as up-and-down the layers of an enterprise. Repeatedly David has earned the respect of and served as a trusted advisor to senior executives at Fortune-ranked companies, including, Lucent Technologies, Southern Company, Rhodia Chemicals, Southern Research Institute, Bell Canada, and others. Before starting Windward in 2000, David founded the Atlanta office of ESI International, serving as the practice leader, assuming full P&L responsibility and leading his office to become ESI's most profitable branch – achieving and maintaining an 80% net margin. A practitioner by choice, David has led teams ranging from 10 to 50 professionals on projects ranging from $10 million to $1 billion in responsibility.

Rodney Turner is a British and New Zealand organizational theorist. He was a Professor of Project Management at Skema Business School from 2004 to 2018 and at the Kingston Business School from 2013 to 2018. He is known for his work of project management. Rodney served as Chairman of the Association for Project Management and as chairman of the International Project Management Association from 1998 to 2000. Turner's research areas cover project management in small to medium enterprises, the management of complex projects, the governance of project management, including ethics and trust, project leadership and Human Resource Management in the project-oriented firm.

J. LeRoy Ward, President of Ward Associates, is a seasoned global executive with more than 40 years' experience in project, program, and portfolio management (PPM) in industry and government. Formerly, Mr. Ward was Executive Vice President of Client Solutions at the International Institute for Learning, and prior to that served as Executive Vice President at ESI International (now Korn Ferry) responsible for global product strategy in PPM, business analysis and leadership. He has broad and deep international business experience, including negotiating licensing partnerships in North America, Europe, Asia, and Australia. He has authored a number of publications and articles including *Project Management Dictionary* (3ʳᵈ. ed); with Ginger Levin, *PMP˚ Exam Practice Test and Study Guide, PgMP˚ Exam Challenge*, and *Program Management Complexity, A Competency Model*; and, with Carl Pritchard, a collection of audio CDs entitled Conversations on Passing the PMP˚ Exam (4ᵗʰ ed), which has helped more than 25,000 professionals earn the credential.

John Wyzalek holds a Doctor of Business Administration in Project Management and Programme Management from SKEMA Business School, France. His research has investigated stakeholder management in project portfolios. He also holds the PfMP certification from the Project Management Institute.

1

The Development of the Project Portfolio Concept

John Wyzalek

This book begins with this chapter that explores the origins of project portfolios in an attempt to explain current conceptions of a project portfolio. Project portfolios are a subject of ongoing research. Recent work has broadened the definition of a project portfolio from that of a collection of projects efficiently managed to meet strategic business objectives (PMI, 2006) to that of a wider theory of the organization (Geraldi, Teerikangas, & Birollo, 2022). In tracing the evolution of the portfolio concept, the chapter examines historical notions as well as the scientific literature to identify conceptual aspects associated with a project portfolio. It analyzes aspects that have stayed constant over time to gain further insight into what is a project portfolio. It also examines aspects that have developed and been elaborated over time. The examination of historical notions begins with a look at how the word portfolio developed in the English language.

ETYMOLOGY AND CONCEPTUAL ASPECTS

This section looks at the current definition and etymology of the word portfolio to establish conceptual facets associated with the word's meaning and historical development. The word portfolio's introduction into English is documented as 1713 (Merriam-Webster.com, 2022), and its origin in English is traced back to the Italian word *portafoglia*, which means to carry (*porta*) a leaf or sheet (*foglia*) of paper (Gove, 1981). *Foglia* is the origin for the word folio, which is defined as a single sheet of paper folded once to create four book pages (Gove, 1981). Folio also embodies a notion

DOI: 10.1201/9781003315902-1

of uniting multiple subcomponents. In 1713, portfolio referred to a case for carrying papers. This definition indicates several conceptual aspects:

- A portfolio is composed of discrete units and unites them.
- A portfolio provides a framework for handling its subcomponents.
- A portfolio, as a carrying case, is associated with movement, is dynamic and is headed to and destination determined by its owner.
- A portfolio's contents can be removed and replaced.

These aspects align with the description of portfolio as a collection of individual projects that are rotated in and out as the portfolio progresses towards a destination defined by its organizational owners (Project Management Institute, 2013, pp. 5–7).

The word took on a new denotation in the early 1800s: "the office and functions of a minister of state or member of a cabinet" (Gove, 1981). This meaning apparently is based on government minister's portfolio holding papers related to the ministry's various functions and responsibilities, as illustrated in this quote from 1835: "The portfolio of the war office was put into the hands of Carnot" (Simpson & Weiner, 1989, XII, p. 153). A ministerial portfolio has an equivalence to a political organization similar to the manner in which a project portfolio represents a business organization. Also, both sense the external environment and reconfigure their composition in response to what they sense.

A third dictionary meaning for portfolio is "the securities held by an investor or the commercial paper held by a bank or other financial house" (Gove, 1981). Simpson and Weiner (1989, XII, p. 153) cite an *Economist* article from 1930 as an early use of this meaning: "This fall is partly due to the bank's failure to secure any of last week's Treasury Bills, which is forcing them to replenish their portfolios in the market." This reference aligns with a project portfolio in several aspects:

- Support for a strategic goal
- Changing composition because of external environmental conditions that increase the risk associated with a strategic goal
- Component diversity
- Risk management

This last aspect plays a significant role in the formalization of the portfolio concept.

A MODERN FORMAL DEFINITION WITH LONG HISTORICAL ROOTS

In 1952, Roy and Markowitz each developed a theory-based definition of a portfolio. The starting point for this refinement is that a diversely composed portfolio can manage risk. The concept of using diversification to manage risk has a long history as Markowitz (1999) cites *The Merchant of Venice* by William Shakespeare:

> *My ventures are not in one bottom trusted,*
> *Nor to one place; nor is my whole estate*
> *Upon the fortune of this present year;*
> *Therefore, my merchandise makes me not sad.*
>
> **(Act I, Scene 1)**

Ong (2014) also cites another early 17th century work, *The School of the Noble and Worthy Science of Defense* (Swetnam, 1617), as an early discussion of diversification as a means to mitigate threats. The following line from Swetnam illustrates this point:

> *He is a fool which will adventure all his goods in one ship, especially if it be in a dangerous voyage, or all his mony* [sic] *at one throw at dice although hee* [sic] *know that the runne* [sic] *of the dice never so well, for he that doth so may hap to loose [sic] all.*
>
> **(p. 56)**

In these early examples, the basic framework of a portfolio is present. A portfolio is a single, unifying structure supported by multiple and diverse component elements. Multiplicity and diversity are used to manage the risk of obtaining a single goal. A risk-based context is also evident. A goods-laden ship is on a dangerous voyage (Swetnam), presumably on the way to a home port to bring riches for the ship's investors.

A portfolio as a collection of securities or investments is a way to manage risk through diversification. The risk and diversification aspects of a portfolio were mathematically formalized by Markowitz (1952) and Roy (1952). Markowitz's theory of portfolio selection assumes the rationality of investors who seek to maximize investment return with a minimum of risk. He mathematically proves that there exists a combination of the highest expected return value with the lowest variance. These pairs of expected return and variance values form "efficient surfaces" (Markowitz, 1952, p. 80). The paper acknowledges that portfolio selection in practice would require statistics and expert knowledge "to form reasonable probability beliefs" (p. 80). The theorem therefore does not minimize the risk posed by relying on expert knowledge, nor does Markowitz (1952) discuss how to keep a portfolio along the efficient surfaces as expert knowledge changes and evolves.

Roy (1952) investigates how a person is expected to know the outcome of a course of action and "the well-known phenomenon of the diversification of resources among a wide range of assets" (p. 431). Roy (1952) explicitly uses the term resources and also explores the portfolio as a mechanism to overcome the risk presented by uncertainty. The paper mentions the role of an investor's psychology and behavior but does not pursue an in-depth investigation. It examines expected return from a group of assets based on past performance. In contrast to Markowitz's efficient surfaces, Roy sets a "disaster" level of return, which is the highest level of unacceptable return or the lowest level of acceptable return. The paper also offers equations on how to structure a portfolio and select components.

CONTRIBUTION FROM ORGANIZATIONAL MANAGEMENT

The 1960s and 1970s saw the development of management strategies that relied on multiple components aligned with goal attainment. The word portfolio was not used except by one such strategic model, but there is great similarity between a project portfolio and these strategic frameworks that use goal-aligned components to maximize the use of resources in order to attain goals. There is also great similarity to efficient surfaces (Markowitz, 1952) and the disaster level of return (Roy, 1952).

Drucker (1963) focuses on efficient use of resources for only those products and services that provide optimal economic return. He argues that the main responsibility of a business manager is: "To strive for the best possible economic results from the resources currently employed or available" (p. 53). This framework proposes prioritizing products, operations and activities in accordance with the strategic goal of best possible economic results. The model also classifies products as "tomorrow's breadwinners," "today's breadwinners," "products capable of becoming net contributors if something drastic is done," "yesterday's breadwinners," "the also-rans" and "the failures" (p. 55). The model states that the first three of these categories should be given priority in resource allocation.

Henderson (1970) introduces the product portfolio and a portfolio-based strategic model, the BCG Growth Share Matrix. Similar to the product classification in Drucker (1963), this model classifies an organization's products into a matrix bounded by growth and market share. It proposes that products with high market share and low growth have a low-cost structure and produce excess revenue that an organization uses to fund high growth, high market share products, which eventually have high market share and low growth. Thus, products follow a lifecycle that sustains an organization. In this model, a portfolio consists of products or the business units that create them, which fit into one of the model's categories that define resource consumption and allocation. BCG Growth Share Matrix is dynamic in that it is structured portfolio of products that changes in response to changes in the external marketplace. The 1970s witnesses the introduction of additional strategic matrix models, including the Shell Directional Policy Matrix (Robinson, Hichens, & Wade, 1978), Arthur D. Little's Strategic Condition Matrix (Udo-Imeh, Edet, & Anani, 2012) and The General Electric Multifactor Portfolio Model, also known as the GE–McKinsey nine-box matrix (McKinsey & Company, 2008).

PROJECT-BASED PORTFOLIOS

The identification of portfolios with product development in the 1960s and 1970s also brings about the identifications of portfolios with projects. Projects are the delivery mechanism for new products. Souder (1988, p. 156) reports that "hundreds of portfolio models have been proposed in the literature." One such model cited by Souder is Dean (1970), which focuses on project selection done quantitatively and systematically. Criteria for project selection include alignment with overall organizational goals as well as research and development, which are considered in the context of available resources to attain these goals. Criteria for project termination include changes in the external environment, resource consumption negatively affecting other portfolio projects, and low probability of achieving technical objects or commercial results. The project portfolio is used to manage risk and to maximize resources and return.

Dean (1968) investigates the evaluation and selection of research and development (R&D) projects, and there is a long association of project portfolios with R&D (Bard, Balachandra, & Kaufman, 1988; Biedenbach & Müller, 2012; Chiu & Gear, 1979; Cooper, Edgett, & Kleinschmidt, 1997a, 1997b; Golabi, 1987; Khorramshahgol & Gousty, 1986; Lechler & Thomas, 2015; Madey & Dean, 1985; Roussel, Saad, & Erickson, 1991; Santiago & Vakili, 2005; Spharim & Szakonyi, 1984). R&D project portfolios feature component projects that are selected mathematically (Bard et al., 1988; Chiu & Gear, 1979; Golabi, 1987; Khorramshahgol & Gousty, 1986; Madey & Dean, 1985; Roussel et al., 1991; Spharim & Szakonyi, 1984) to ensure efficient use of resources and to manage risk.

McFarlan (1981) emphasizes the portfolio as a framework for managing risk in information technology projects, furthering associating a project portfolio with innovation. Risk in McFarlan is the threat of project failure in terms of delivering an information system that is incompatible with other organizational systems, schedule and budget overruns, and an inability to deliver a functioning system.

A project portfolio whose components are strategically aligned manages risk by maximizing value and efficient use of resources. The identification of a project portfolio with organization value was strengthened in the 1990s and early 2000s. Cooper et al. (1997a, 1997b) report on portfolio models that link projects with strategy and maximize value. A project portfolio is aligned with "business objectives, combining performances of its components in order to maximize the shareholder's value while balancing resource allocation and risks" (Constantino, Di Gravio, & Nonino, 2015, p. 1744). It is therefore an investment vehicle that maximizes value in a risky environment. Killen, Hunt and Kleinschmidt (2008) also affirm a portfolio as a framework for maximizing value for a given level risk because it enables decision-makers to balance projects with long- and short-term goals and efficiently allocate resources. Kaiser, El Arbi and Ahlemann (2015) also report on how a portfolio enables an organization to respond to market risks and strategically realign and restructure itself in order to maximize value. Marnewick (2018) proposes that a project portfolio and associated work can aid organizational managers to sustain the delivery of value.

More recent research shifts the focus from a portfolio as defined by its components to a portfolio in relation to its environment. Risk management is present in research on how a project portfolio can sense the external environment to identify threats and opportunities (Biedenbach & Müller, 2012; Killen & Hunt, 2010; Killen, Jugdev, Drouin, & Petit, 2012; Petit & Hobbs, 2012; Patanakul, Curtis & Koppel, 2013). Killen and Hunt (2010) apply dynamic capability theory (Teece, Pisano, & Shuen, 1997; Teece, 2009) to the management of project portfolio. In part, they revisit the project selection problem, previously addressed by ensuring internal strategic alignment, by addressing how a project portfolio is aligned strategically in relation to the external environment. This alignment is facilitated by sensing the external environment to develop a strategic response to this environment. Petit and Hobbs (2012) and Kaiser et al. (2015) present empirical research on project portfolios in the context of a dynamic external environment.

Recent research broadens the concept of a portfolio as a framework for managing multiple projects supporting organizational goals. Martinsuo and Geraldi (2020) define a project portfolio as an organization in their investigation of the portfolio in its internal and external context. Vedel and Geraldi (2020) abstract the concept's strategic aspects further to the point where project portfolios "manifest strategy" and "represent an organisation's investments in the future" (p. 255). Geraldi, Teerikangas and Birollo (2022) see a portfolio as a way of organizing project work indicating a path towards a project-based theory of the firm in which a project portfolio shapes and creates a firm.

DISCUSSION AND CONCLUSION

A through line in the development of the project portfolio concept is its association with strategy. Early portfolio models for R&D projects (e.g., Dean, 1968) strategically align with organizational goals for new product development, and recent research equates a project portfolio with strategy (Vedel & Geraldi, 2020). Closely associated with the strategic aspect is risk management. R&D project portfolio models investigate project selection that efficiently uses resources to manage risk. Later conceptions of the project portfolio (Killen et al., 2008) manage risk by maximizing value to an organization.

The portfolio as a framework to manage risk is a strong historical through line running through the financial portfolio models of Markowitz (1952) and Roy (1952), which also seek to maximize value. Markowitz and Roy emphasize managing risk through a diversification of portfolio components, and diversification, as Markowitz, 1999 points out has a history stretching for hundreds of years. Another historical connection is associating a portfolio with an organization and its functions as noted in an 1835 reference to the portfolio of the war office (Simpson & Weiner, 1989) and as investigated in research on the project portfolio related to management theory of the firm (Killen et al., 2012; Geraldi et al., 2022).

Perhaps the strongest and oldest historical through line is the portfolio as a framework for organizing and unifying its multiple components. The word portfolio originated in English to describe a case for carrying sheets of paper (Gove, 1981). It then was adapted to mean an office of state and its function and responsibilities. A further meaning was developed from the notion of a carrying case of papers to a financial portfolio holding multiple types of paper, where the paper is a financial instrument. The meaning that is important in business and project management is a portfolio as a collection of multiple products or projects (Dean, 1970). Like a carrying case that stays around as papers are placed in and out of it, project portfolio components can be changed, but the framework of the portfolio remains (Marnewick, 2018). A portfolio's permanency is enabled by the ability to change portfolio components (Killen & Hunt, 2010; Patanakul et al., 2013; Petit & Hobbs, 2012). This sustained dynamic is what makes a project portfolio adaptable and should ensure its permanency in the field of project and organizational management.

The following chapters in this book are mainly concerned with portfolio management, but they touch on themes explored in this chapter. For example, Killen (Chapter 12) discusses dynamic capability; Tennant (Chapter 2) examines strategy and new product development; Bourne (Chapter 10) looks at delivering value; and Petit, Romero-Torres and Delisle (Chapter 11) examine strategic alignment in agile portfolio management. The remainder of the book looks at portfolio governance (Chapter 4 by Ward and Chapter 5 by Baker), and the portfolio in its external context, whether at the organizational or market level, and examines issues not covered in this chapter. Aubrée-Dauchez (Chapter 8), Curlee (Chapter 9) and Turner (Chapter 6) each look at the role of communications in portfolio management while Baker (Chapter 3) examines establishing a new portfolio and Marnewick (Chapter 7) discusses successful project portfolio management.

REFERENCES

Bard, J. F., Balachandra, R., & Kaufmann, P. E. (1988). An interactive approach to R&D project selection and termination. *IEEE Transactions on Engineering Management, 35*(3), 139–146.

Biedenbach, T., & Müller, R. (2012). Absorptive, innovative and adaptive capabilities and their impact on project and project portfolio performance. *International Journal of Project Management, 30*(5), 621–635.

Chiu, L. K., & Gear, T. E. (1979). An application and case history of a dynamic R&D portfolio selection model. *IEEE Transactions on Engineering Management, 26*(1), 2–7.

Constantino, F., Di Gravio, G., & Nonino, F. (2015). Project selection in project portfolio management: An artificial neural network model based on critical success factors. *International Journal of Project Management 33*(8), 1744–1754.

Cooper, R. G., Edgett, S., & Kleinschmidt, E. (1997a). Portfolio management in new product development: Lessons from the leaders-I. *Research Technology Management 40*(5), 16–28.

Cooper, R. G., Edgett, S., & Kleinschmidt, E. (1997b). Portfolio management in new product development: Lessons from the leaders-II. *Research Technology Management 40*(6), 43–52.

Cooper, R. G., Edgett, S. J., & Kleinschmidt, E. J. (1999). New product portfolio management: Practices and performance. *The Journal of Product Innovation Management, 16*(4), 333–351.

Dean, B. V. (1968). *Evaluating, selecting, and controlling R&D projects.* New York: American Management Association, Inc.

Dean, B. V. (1970). *Project evaluation: Methods and procedures.* New York: American Management Association, Inc.

Geraldi, J., Teerikangas, S., & Birollo, G. (2022). Project, program and portfolio management as modes of organizing: Theorising at the intersection between mergers and acquisitions and project studies. *International Journal of Project Management, 40*(4), 439–453.

Golabi, K. (1987). Selecting a group of dissimilar projects for funding. *IEEE Transactions on Engineering Management, 34*(3), 138–145.

Gove, P. B. (Ed.). (1981). *Webster's third new international dictionary of the English language, unabridged.* Springfield, MA: Merriam-Webster.

Henderson, B. (1970). *The product portfolio.* Boston: Boston Consulting Group. Retrieved from https://www.bcgperspectives.com/content/Classics/strategy_the_product_portfolio/

Kaiser, M. G., El Arbi, F., & Ahlemann, F. (2015). Successful project portfolio management beyond selection techniques: Understanding the role of structural alignment. *International Journal of Project Management, 33*(1), 126–139.

Khorramshahgol, R., & Gousty, Y. (1986). Technical management notes: Delphic Goal Programming (DGP): A multi-objective cost/benefit approach to R&D portfolio analysis. *IEEE Transactions on Engineering Management, 33*(3), 172–175.

Killen, C. P., & Hunt, R. A. (2010). Dynamic capability through project portfolio management in service and manufacturing industries. *International Journal of Managing Projects in Business, 3*(1), 157–169.

Killen, C. P., Hunt, R. A., & Kleinschmidt, E. J. (2008). Project portfolio management for product innovation. *International Journal of Quality & Reliability Management, 25*(1), 24–38.

Killen, C. P., Jugdev, K., Drouin, N., & Petit, Y. (2012). Advancing project and portfolio research: Applying strategic management theories. *International Journal of Project Management, 30*(5), 525–538.

Lechler, T. G., & Thomas, J. L. (2015). Examining new product development project termination decision quality at the portfolio level: Consequences of dysfunctional executive advocacy. *International Journal of Project Management 33*(7), 1452–1463.

Madey, G. R., & Dean, B. V. (1985). Strategic planning for investment in R&D using decision analysis and mathematical programming. *IEEE Transactions on Engineering Management, 32*(2), 84–90.

Markowitz, H. (1952). Portfolio selection. *The Journal of Finance 7*(1), 77–91.

Markowitz, H. M. (1999). The early history of portfolio theory: 1600–1960. *Financial Analysts Journal, 55*(4), 5–16.

Marnewick, C. (2018) *Realizing strategy through projects: The executive's guide*. New York: Taylor & Francis Group.

Martinsuo, M., & Geraldi, J. (2020). Management of project portfolios: Relationships of project portfolios with their contexts. *International Journal of Project Management, 38*(7), 441–453.

McFarlan, F. W. (1981). Portfolio approach to information-systems. *Harvard Business Review, 59*(5), 142–150.

McKinsey & Company (2008). Enduring ideas: The GE-McKinsey nine box matrix. *McKinsey Quarterly, 45*(3). Retrieved from http://www.mckinsey.com/business-functions/strategy-and-corporate-finance/our-insights/enduring-ideas-the-ge-and-mckinsey-nine-box-matrix

Merriam-Webster.com. Retrieved October 8, 2022, from https://www.merriam-webster.com/dictionary/portfolio

Ong, F. (2014). What is the origin of the saying 'Don't put all your eggs in one basket'? Retrieved from https://www.quora.com/What-is-the-origin-of-the-saying-dont-put-all-your-eggs-in-one-basket

Patanakul, P., Curtis, A., & Koppel, B. (2013). *Effectiveness in project portfolio management*. Newtown Square, PA: Project Management Institute.

Patel, P., & Younger, M. (1978). A frame of reference for strategy development. *Long Range Planning, 11*(2), 6–12.

Petit, Y., & Hobbs, B. (2012). *Project portfolios in dynamic environments: Organizing for uncertainty*. Newtown Square, PA: Project Management Institute.

Project Management Institute (2006). *The standard for portfolio management*. Newtown Square, PA: Project Management Institute.

Project Management Institute (2017). *A guide to the project management body of knowledge (PMBOK® guide)*, 6th ed. Newtown Square, PA: Project Management Institute.

Robinson, S. J. Q., Hichens, R. E., & Wade, D. P. (1978). The directional policy matrix—Tool for strategic planning. *Long Range Planning, 11*(3), 8–15.

Roussel, P. A., Saad, K. N., & Erickson, T. J. (1991). *Third generation R&D: Managing the link to corporate strategy*. Boston, MA: Harvard Business Review Press.

Roy, A. D. (1952). Safety first and the holding of assets. *Econometric 20*(3), 431–449.

Santiago, L. P., & Vakili, P. (2005, July). Optimal project selection and budget allocation for R&D portfolios. In T. R. Anderson, T. U. Daim, D. F. Kocaoglu, D. Z. Milosevic, & C. M. Weber (eds.), *Technology Management: A Unifying Discipline for Melting the Boundaries* (pp. 275–281). Portland, OR: PICMET.

Simpson, J. A. & Weiner, E. S. C. (Eds.). (1989). *The Oxford English Dictionary*, 2nd ed. (vol XII). Oxford, UK: Clarendon Press.

Souder, W. E. (1988). Selecting projects that maximize profits. In D. I. Cleland & W. R. King (Eds.). *Project management handbook*, 2nd ed. New York: John Wiley & Sons.

Spharim, I., & Szakonyi, R. (1984). A simple method for evaluation and selection of R&D proposals for a competitive grant fund. *IEEE Transactions on Engineering Management*, *31*(4), 184–185.

Swetnam, J. (1617). *The school of the noble and worthy science of defense*. Retrieved from http://www.aemma.org/onlineResources/swetnam/joseph_swetnam.pdf

Teece, D. J. (2009). *Dynamic capabilities and strategic management: Organizing for innovation and growth*. Oxford, UK: Oxford University Press.

Teece, D. J., Pisano, G., & Shuen, A. (1997). Dynamic capabilities and strategic management. *Strategic Management Journal*, *18*(7), 509–533.

Udo-Imeh, P. T., Edet, W. E., & Anani, R. B. (2012). Portfolio analysis models: A review. *European Journal of Business and Management*, *4*(18), 101–120.

Vedel, J. B., & Geraldi, J. (2020). A 'stranger' in the making of strategy: A process perspective of project portfolio management in a pharmaceutical firm. *International Journal of Project Management*, *38*(7), 454–463.

2

Corporate Strategy and Project Management for Developing a Portfolio of Products

David Tennant

INTRODUCTION

In developing new products, there are many moving parts to consider: Who will lead the effort? Does the new product (s) line up with the company's strategic plan? What are the roles of the various functional areas? How will we know if it will be successful? How do we manage multiple ongoing projects/products?

There will usually be many different departments involved in a new product's development. This chapter will evaluate, at a high level, the integration of the corporation's focus, the role of various departments involved, and techniques to plan, design, and launch the new product.

WHAT IS A PRODUCT?

A product can be an item or service that provides value to customers for a price. It is easy to think of products simply as new HDTV or a new smart phone. However, products can also be a service. Examples could include the electric service you get from your utility or the ISP that provides your home with internet access. Or the accounting firm that your company uses for accounting and taxes. The line can become further blurred if we think about an app for your smartphone. Is this a service or a firm product?

DOI: 10.1201/9781003315902-2

Regardless, hard products are marketed and sold differently from services. Further, how are specific services or products developed? At what point is the decision made to develop, or improve, an existing product? For large companies, this is forever an objective to have the right portfolio of products and services that serve to attract new customers while retaining existing ones.

WHERE DO THE IDEAS FOR NEW PRODUCTS COME FROM?

What is strategic planning? Why is it important? Strategy has different meanings to different people. To many, a strategy represents a guiding philosophy. To others, it may be a roadmap to take a company from point A to point B.

From a practical standpoint, a strategic plan is a long-term vision for the corporation. As an example, a company's financial objective may be to increase revenues by 20% each year for the next five years. How would this be accomplished? A strategic plan would delineate the tasks to make this happen. Examples might include the following:

- Develop five new related products
- Acquire a firm with complimentary products and services in the next 24 months (Growth through acquisition)
- Improve and enhance the current product lineup
- Increase spending on R&D (Research and Development)

The strategic plan should typically look forward around five years. Note that the mentioned tasks would require significant investment, detail, and planning.

For large corporations, the marketing group plays a large role in the strategic planning process. This is because marketing is concerned with industry trends, customer preferences, competitor analysis, and with developing promotions to keep the company's brand alive and relevant. Therefore, it should be clear that a professional marketing organization is needed to drive parts of the strategic plan.

To develop an effective strategic plan, there will be a many "inputs."

- Purpose of the company (mission)
- Objectives
- Where do we want to be in 5 years (or 10 years)?
- Industry trends
- Key stakeholders (current and future)
- Competitor analysis
- Strategic initiatives
  1.
  2.
  3.
- Estimate to develop the product or service
- Schedule (when will it happen?)
- Risks associated with the initiatives
- Key resources needed (dollars, people, equipment, etc.)
- Metrics (How will we know when we get there?)

A true industry leader needs to anticipate the market and plan accordingly.

Signs of market leadership include:

- Your competitors consider your firm to be innovative, a rule maker (or breaker) and nimble in coming to market.
- Top management are not happy with the status quo but are always seeking to do things differently (i.e., better).
- The firm is focused on creative innovation. Operational efficiency is a secondary priority (this may require investment in R&D).
- Senior management is focused on strategy; middle management is focused on process.
- Your customers look forward to your new products and you have a reputation for quality (Key point: People will pay more for higher quality, or even the perception of higher quality).

IMPLEMENTING THE STRATEGIC PLAN

In many instances, launching the strategic plan involves using the organization's many departments and resources in the form of projects. Note that projects are meant to support specific objectives and usually a strategy involves multiple objectives. In this medium, it is appropriate to note that project management can – and must – play a critical role in planning and executing the projects or products associated with the new strategic direction.

The discipline of project management is well suited to derive projects from the strategic plan. Note that the development of products can be considered as a project.

The principles of project management can be utilized to ensure that strategic initiatives are planned and implemented within budget, schedule, and quality parameters. To a large degree, project plans are very similar to the key components of a strategic plan.

It is interesting to note that the project plan represents a "drill down" from the strategic plan. Further, if a project plan does not support the strategic plan, it should not be considered or funded.

Table 2.1 compares the components of a strategic plan with a project plan.

There are many considerations in implementing the strategic plan. For example, will this require a culture change? Do we have the right people, internally, to implement the plan?

Communicating the plan is important. To achieve buy-in, people need to understand why such changes are occurring. It is incumbent on senior

TABLE 2.1

Similarities of Strategic and Project Plan Elements

Strategic Planning Elements (High-Level Perspective)	Project Planning: Product Development (Detailed Project-Level Perspective)
1. Goals and purpose	1. Objectives
2. Timeline	2. Detailed schedule with milestones
3. High-level budget	3. Detailed budget with contingency
4. NPV and ROI[a] Analysis	4. Requirements and equipment resources
5. Resources needed	5. Communications management
6. Communicating the vision	6. Client acceptance criteria (Client: CEO)
7. Metrics	7. Supply chain management
8. Risk analysis[b]	8. Risk review and analysis
9. Culture change	9. Quality management
10. Competitor analysis	10. Stakeholder management
11. Market trends	11. Project integration
12. Specific project ssignments	12. Scope management

a NPV = Net Present Value (a form of cost–benefit analysis); ROI = Return on Investment
b Many companies do not perform risk reviews with their strategic planning; however, the author believes this is an important component and should be included.

management (typically the CEO) to outline why the change is needed, when it will go into effect, the benefits that will accrue to the company and its employees, and how this will occur. It is necessary to put together a project plan (or similarly, an implementation plan) to move strategies forward.

PUTTING STRATEGY INTO ACTION

Putting strategic initiatives into motion is where the discipline of project management can assist greatly. Consider the following examples:

- All the world's car manufacturers are moving away from the internal combustion engine to electric vehicles.
- Electric utilities are moving away from coal-fired power plants to clean energy such as wind, solar, and natural gas.
- Robotics are seeing significant use in manufacturing to cut costs and to manufacture products of consistent quality.
- The Internet of Things – Smart Cities. A smart city can connect with cloud-based applications to improve city services: reduced traffic congestion, "connecting" cars for collision avoidance, and helping drivers find available parking in congested areas.

It should be apparent that a strategic plan must be concise and clearly articulated. In each of the given scenarios, project management will be needed to ensure strong planning and successful implementation.

All the given examples are complicated and capital-intensive proposals. How would a company plan and organize these strategic initiatives? Let's consider the roles of key functional areas in product development.

ROLES OF FUNCTIONAL GROUPS IN PRODUCT DEVELOPMENT

Most new product development is led by either the marketing or engineering department. Depending on the product or the firm's organizational structure, these two groups generally have the best feel for the product's features and customer wants.

Marketing

As noted previously, the marketing department has a responsibility to keep abreast of customer preferences, industry trends, and competitor analysis (including their products, pricing, quality, etc.). However, the marketing group *cannot* predict a product's success, profitability, or customer acceptance. So, how does this help?

During product development, marketing will generally perform customer focus groups. This will gage how customers react to the new products, their likes or dislikes, suggestions for features, etc. Focus groups are a strong tool to assist marketing to communicate to the engineering group suggestions for product design – to better gain market acceptance. Focus groups can also suggest how much customers are willing to pay for a new product.

Marketing will also develop cost and pricing models (usually an excel spreadsheet) that will illustrate a number of "what if" scenarios. Table 2.2 is an example of sensitivity analysis based on price and profitability vs. number of units sold.

This table is based on several assumptions. It can be assumed that the cost per unit decreases as more units are produced (economies of scale). And there are assumptions about how many units can be sold and what the profit margin would be based on various pricing scenarios.

To determine the cost of the new product, input is required from both the engineering and manufacturing departments. Marketing is responsible for the new product's pricing, promotion, and distribution.

TABLE 2.2

Price and Profitability Sensitivity Analysis

Units Sold	Price	TTL Revenue	Cost/Unit	TTL Costs	Margin
500,000	$ 40	$ 20,000,000	$ 15	$ 7,500,000	$ 12,500,000
500,000	$ 30	$ 15,000,000	$ 15	$ 7,500,000	$ 7,500,000
500,000	$ 20	$ 10,000,000	$ 15	$ 7,500,000	$ 2,500,000
600,000	$ 40	$ 24,000,000	$ 13	$ 7,800,000	$ 16,200,000
600,000	$ 30	$ 18,000,000	$ 13	$ 7,800,000	$ 10,200,000
600,000	$ 20	$ 12,000,000	$ 13	$ 7,800,000	$ 4,200,000
750,000	$ 40	$ 30,000,000	$ 9	$ 6,750,000	$ 23,250,000
750,000	$ 30	$ 22,500,000	$ 9	$ 6,750,000	$ 15,750,000
750,000	$ 20	$ 15,000,000	$ 9	$ 6,750,000	$ 8,250,000

(Table developed by David Tennant)

Engineering

Engineering plays a major role in the design and testing of the new product. This will include determining the materials, lifespan, design for misuse, and ergonomics.

The selection of materials will determine the strength (and cost) of the product. For example, to make a bicycle suitable for a 200 lb. person, we would design the bike to support someone who weighs up to 250 lb. for a safety factor of 25%. This could have an influence on the type of materials: aluminum vs. steel alloy; and the tube thickness of the frame, 1/8" thick vs. 3/16". Engineering design in the past used to develop six or eight prototypes which would then be tested. If each prototype failed, the design would be revised and new prototypes produced for another round of tests. Today, engineering teams use sophisticated and specialized modeling software that can reduce design time and result in fewer prototypes (e.g., one or two prototypes instead of six). The modeling does not eliminate the need for prototypes or testing, but it does allow for more efficient design, more precise analysis, higher quality, and fewer test units. This saves both time and money and allows products to come to market more quickly.

Engineers must also consider the misuse – whether by accident or intentional – when designing a new product. This is because we live in a litigious society and people may sue manufacturers if they are injured using your product. As an example, people may step onto the very last (high) step on a ladder even though warning labels caution against this. Similarly, they may place the ladder at a dangerous angle to a wall. Engineers must plan for their products being misused and attempt to prevent this from happening through robust design.

Ergonomics, also known as "human factors," is another consideration. This is the study and design of products that will easily fit 90% of the population. For example, if you design a new office chair (on wheels), it should be designed so that 90% of the population can comfortably sit on it: tall people, short people, different ages, weights, etc. If you are designing a car interior, can the driver easily see and manipulate the displays and controls? Can the driver's seat be adjusted to fit 90% of the population?

It should be noted that there will generally be some tension between the marketing and engineering teams. This is because the two groups many times have differing perspectives. The engineering team is concerned with designing a safe and attractive product within a given schedule and budget. Changes should be kept to a minimum once design has reached a specific point.

Marketing, on the other hand, is always monitoring customer trends and tastes. When market dynamics change, it may be useful to request changes to the product. This can impact the design, materials, costs, and schedule etc. Therefore, there will be tension in balancing the "wants" of customers vs. the progress of engineering design.

It should be noted that software products have traditionally run over budget and schedule during development and implementation due to unforeseen problems and customer changes. However, new project management techniques have been developed for software. Many companies have shifted to an "Agile" design philosophy whereby a product is developed using mini projects (called "sprints") with a different approach to the product's development. The reader is encouraged to study further on this topic if it applies to you.

OTHER ORGANIZATIONS INVOLVED WITH PRODUCT DEVELOPMENT

Depending on the importance, capital investment, and complexity, here are other groups that may be involved with your product team:

Legal – A legal review of the new product may be needed to determine the prospects for misuse and any legal warnings (labels) that may be required. Also, if the new product will be patented, a patent attorney will need to be involved.

Manufacturing – The role of manufacturing is to work with the engineering team to determine the most efficient production methods and to minimize scrap materials (quality control). If using robotics, it is likely some programming of the machines will be needed. To confirm quality, a number of random products may be selected for testing (5%? 10%?).

Supply Chain – This group is also known as procurement in some organizations. It is concerned with issuing RFPs, contracts, and purchase orders.

Finance – The CFOs office closely watches the budget, calculating the NPV and ROI (net present value and return on investment) as the product progresses through development.

It is appropriate to note that projects (products) must be careful to stay within the budget and schedule. If costs spiral out of control, the new product may become unprofitable and never recover the development costs. There have been many products (and companies) that failed due to cost overruns.

CHANGES TO THE STRATEGIC PLAN

Staying focused on the objectives and strategy is important. However, there are times when the strategy may need to change and, in turn, the supporting projects or products. It is necessary to exhibit flexibility.

There are key tipping points when a company's strategy can be upset. These may include the following:

- A merger or acquisition
- Transition from a publicly held to a privately held position (or vice-versa)
- Change in top leadership (different priorities)
- Changes in federal or state regulations
- Changes in market conditions

As product development progresses, it is important that key milestone dates are met and budgets are followed. There are several ways to track a project's progress, and these are common to the project management profession. These can include using earned-value calculations, placing key milestones within a project schedule, and detailing project cash flows to some degree of certainty.

PROJECT RISK REVIEWS

It is important to emphasize that project risk reviews, performed on a regular basis, can prevent problems from developing or provide enough advance notice so that minimal disruption occurs. Project risk reviews are a strong management and problem-prevention tool. Many times, project managers complain that they are constantly putting out fires. Risk reviews generally eliminate the need for constant firefighting.

Every project should have a formal risk review on a recurring basis. This is a team exercise that attempts, through brainstorming, to identify potential future problems. Once problems have been identified, strategies can be developed to minimize or prevent them from occurring.

A risk review is not a one-time event. Risk reviews should occur throughout the project's lifecycle (initiation, planning, execution, and closing) on a regular basis. The advantages of risk reviews include the following:

- Proactively acting today to prevent chaos later
- Reducing or eliminating "firefighting"
- Thinking ahead to develop strategies to prevent problems
- Helping to define the scope, costs, and schedule of the project
- Assisting the team in developing a contingency budget
- Assisting the team in developing their management skills

To prevent further problems from developing, it is important that all stakeholders are identified during a project's conception. This is another way to mitigate changes and/or risks. A stakeholder is a person (or group) that has a vested interest in the project. Some stakeholders are more important than others. For example, a senior executive may want to provide input and ideas for the new product. If this person is not identified early, this is guaranteed to create problems later. The important stakeholders want to provide input and, generally, one of them will be supporting the project financially. This is referred to as the sponsoring executive or sponsoring stakeholder. It is important that the project leader identifies as many stakeholders as possible. Note that stakeholders can also include entities external to the company such as key suppliers or regulatory agencies (Federal or State).

Table 2.3 illustrates a qualitative risk review for the development of a new generation of solar panels. P refers to the probability of the risk occurring, and I is the impact should the risk occur. H, M, and L refer to whether to risk has a high, medium, or low probability of occurring, and a high, medium, or low impact of the risk does occur. These rankings are determined by the team when performing the risk review.

TABLE 2.3

Risk Review for New Solar Panel Product

Risk	P	I	Strategies
Solar panel materials delivered late	M	M	- Put penalties in place for late delivery - Select reliable supplier(s) - Place orders early
Due to inflation, costs exceed 20%	H	M	- Seek cheaper materials - Negotiate with suppliers
Quality issues with panels	L	H	- Ensure OA with regular design reviews - Confirm suppliers QA/QC programs - Perform adequate panel testing during prototype development
Panel fails initial product testing	L	H	- Ensure QA with regular design reviews - Confirm suppliers QA/QC programs - Perform adequate panel testing during prototype development
Production unable to keep up with demand	M	M	- Work with suppliers to ensure adequate delivery - Contract with outside manufactures for overflow work - Work with manufacturing group to add additional shifts or efficiencies
Competitors product is cheaper	M	M	- Negotiate reduced costs with suppliers - Shift production to more efficient facility - Provide a higher quality and superior product

(Table developed by David Tennant)

CONNECTING THE DOTS

The strategic plan lays out the corporation's vision and roadmap for the future. The output will always be new products, services, and projects that support the corporate vision. And recall that the development of products and services are projects. For a large company, this can translate in multiple "projects" (i.e., products) under simultaneous development. How do we therefore manage a portfolio of moving projects? There are some strategies to assist with directing multiple projects:

- Each project should have a project/implementation plan. If several of the projects are similar, it may be useful to manage them together as a program (this is especially true if the same design team or supplier is involved with several of the projects).
- Some companies implement a Project Management Office (PMO) to oversee or support all projects. The PMO can be a temporary or permanent entity.

A program is group of related projects managed in a coordinated way to obtain benefits and control not available by managing them individually. Program management focuses on project interdependencies:

- Resolving resource constraints
- Aligning organizational or strategic direction
- Resolving issues and change management

Figure 2.1 illustrates how a portfolio of programs and projects can be organized.

Depending on the project's complexity, it may be useful to develop a project organization chart. Figure 2.2 is a sample project organization chart.

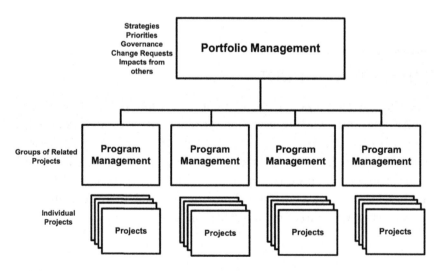

FIGURE 2.1
Projects, programs, and portfolios. (Figure developed by David Tennant).

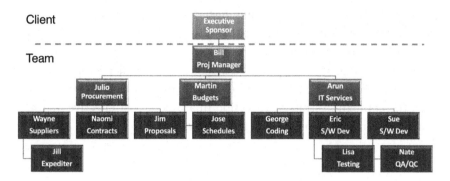

FIGURE 2.2
Sample project organization chart. (Figure developed by David Tennant).

SUMMARY

Project management provides an excellent vehicle for the implementation of strategic initiatives. A firm's strategic plan will generally focus on market conditions in the future and devise a strategy to adjust to these conditions. The senior management team performs strategic planning with inputs from various departments – especially marketing.

The strategic plan components are very similar to those of a project plan; hence the hand-off to project managers should be relatively easy.

There must be some flexibility built into both the strategic and project plans. Due to changes in market trends, regulatory issues, or other actions, it may be necessary to change plans midstream. If flexibility is built into the planning stages, this can be dealt with.

Communicating both the strategic and project plans is highly desired. It has been the author's experience that poor communications are one of the issues associated with project failure. It is the role of the CEO to clearly articulate the strategic plan's vision and purpose.

Product development will always involve multiple departments and stakeholders. The marketing group is especially important as this group typically drives the new product's development. They are responsible for recognizing industry trends, customer preferences, and developing pricing/cost models.

Other key product development stakeholders include Finance, Engineering, Legal and Supply Chain departments. It is important that schedules and budgets are well defined and adhered to during a product's development. Failure to maintain budgets and schedules can lead to product cancellation as development costs cannot be recovered or the product will not be competitive in the marketplace.

In summary, products or projects will be an output of a company's strategic plan. And well-planned and executed projects will effectively support the company's strategic direction in a timely, cost-effective manner, while reinforcing the purpose of the corporation.

3

Developing a New Portfolio

Jennifer Young Baker

For a majority of practitioners, the portfolio that they work on is an existing portfolio with expected goals and objectives. Creating a new portfolio requires some additional steps and direction to ensure future success. There are four general areas of focus within the portfolio management space – strategic alignment, governance, execution/delivery, and methods/tools. Before any of this can happen, there has to be consensus regarding the need for change. One of the most significant components of this portfolio's creation is in the area of change management embedded into the design and delivery of the portfolio.

Why change management? Loosely defined, change management is the way an organization describes and implements changes in its environment including preparing and supporting their teams, establishing what is necessary for the change to occur, and then monitoring its implementation activities to ensure success. When you look at what a portfolio does and its scope, it is easy to see the type of communication and training that would be needed – particularly in a company that has never had a defined portfolio.

Before a portfolio charter can be created, there needs to be a common understanding and consensus that this step is needed. Within an organization that has never had a defined portfolio, this may be difficult to achieve. In an immature environment which is common in this scenario, there is significant pushback or change resistance with the "we've always done it this way" approach for some members of management. While that approach may have been successful in the past, the mere fact that the creation of a portfolio is a discussion implies growth has occurred. Albert Einstein once said, "The measure of intelligence is the ability to change"

DOI: 10.1201/9781003315902-3

(Partridge, 2020). The portfolio manager's most significant task at this point is to facilitate understanding and consensus.

If the creation of the portfolio is forced without the consensus and understanding needed, there is a greater likelihood of failure (Schibi, 2013). When looking at why portfolios fail or do not deliver as expected, there are many reasons that point back to its creation, such as:

- being set up for the wrong reasons,
- organizational acceptance is missing or only partially present,
- having unrealistic expectations of what portfolios can and cannot do within the environment, and
- incorrect or unsuitable setup.

Portfolio components should be fulfilling strategic goals, so it makes sense to look at the strategic planning process to determine how consensus is determined for lessons and leadership to proceed. Typically, team members responsible for managing value-added activities within key areas in the company actively participate in the process because the implementation of strategy is not the sole jurisdiction of the CEO and the senior executive team (GPG, 2019). Simon Sinek famously discusses "Find Your Why" and helping teammates to understand why helps establish understanding and in some cases a sense of ownership (Sinek, 2020). To further emphasize this point, CIO author Stephanie Overby writes to leaders, "It's not about your vision; it's about their why" (Overby, 2022). The portfolio should focus on maximizing the value of the investment decisions that are made. In order to do this, the portfolio leadership needs to be integrated with the strategic planning process to establish the portfolio roadmap.

The team members executing the work sharing information about operations and execution enable senior executive team members to ascertain if the plans in mind are reasonable and realistic to deliver. Additionally, if the teams executing the work feel engaged in the process, they develop a sense of ownership. These team members should be supported by being provided with education and training needed to better execute the process. This learning component should not only apply to teams to build new skills but leadership as well. Once everyone has the skills, knowledge, and perspective, it is easier to build enthusiasm among the collective team members both individually and collectively (GPG, 2019).

One crucial component of the portfolio startup process is to identify the stakeholders and the value that is delivered. This is everyone who has something to gain or lose with the implementation of this portfolio. Why lose? You may be asking; shouldn't this be a win–win scenario? In an idealistic sense, yes it should. However realistically, someone who had control of what gets done is losing some or all of that control. This is where the value component kicks in. There is a problem or an opportunity to be solved with the portfolio's work and that solution provides value to the organization. Some stakeholders will help to define the problem; some will help find potential solutions; others will implement those solutions; and others will use that solution while the organization reaps the benefits or value that those solutions provided. Each group of stakeholders has a role to play and has needs that should be addressed throughout the process (PMI, 2017). Knowing who the winners and losers are in the situation allows the portfolio leadership to employ tactics to support that individual stakeholder where they are and ultimately deliver the portfolio support needed. These stakeholders will become part of the governance process for the portfolio, so it is very important to understand not only where they are coming from but also how to meet their needs.

When a new portfolio is established, the governance process which supports it needs to be simple and straightforward. Early complexity makes for unnecessary difficulties for both the stakeholders and the practitioners. Remember that "...governance models are simply the way the organization chooses to apply ...governance. It covers the roles involved in decision-making processes and the processes, policies and 'internal rules' ... the approach to managing, controlling, and reporting on the work." (Ten, 2021) One place where you may want to instill flexibility is with tiered rulesets based on the complexity of the project or program. This doesn't need to be difficult (simpler projects – simpler rules, more complex – more rules, most complex – most rules). Sponsors and steering committees should be helping with the governance of projects and programs. This will mature over time.

Understanding where the maturity of the organization exists allows the portfolio leadership to understand how to present information. This assessment can be through a formal process or something more akin to a pulse check type of analysis. This information is helpful not only with the definition period for the portfolio but also in defining portfolio functions, the portfolio roadmap and encapsulates the findings into the implementation plan activities. Even if you think that your organization is more mature, target the lowest common denominator. Why?

According to Gartner, around 80% of PMO's are level 1 or 2 maturity. Let that sink in. Unless you know you are an elite PMO, you are part of the 80%. In fact, without actively maturing your organization's portfolio processes, you will remain in that 80% for many years. Organizations simply don't "get there" by the passage of time; it takes active work to achieve desired improvement targets to arrive at better portfolio management processes.

(Washington, 2019)

There has to be a clear link between the executed components and delivery of the strategy/strategic elements. This is framed in the portfolio charter and then enhanced through the portfolio plan. That plan identifies the stakeholders, their role and how they are engaged throughout the process. This is not only the communication style and mechanism, but the plan should also reflect their level of enthusiasm and support as well as the training needed. Knowing this along with how the portfolio value is interpreted by each stakeholder makes the communication easier by targeting those discussions to answer their "why" or to use some acronyms – What's In It For Me (WIIFM). Remember that value is contextual – different perspectives from each stakeholder.

Initial portfolio reports should be simple. Choose a few KPIs that are relevant to the strategic goals and objectives and speak to the project and program delivery underway. Structure the comments or the narrative for those KPIs to the individual stakeholders by presenting the information in such a way that it shows the benefit and value which speaks to that stakeholder or stakeholder group. Focus on objective business goals – 37% of executives report that the primary reason for organizational failure is the lack of clearly defined goals and objectives without necessary targets and milestones to measure progress (Ecosys, 2019).

According to PMI,

All portfolios are managed to enhance and maintain the value of the organization, whether that value is tangible or intangible. In order for value to be maximized, the organization should adhere to a set of principles to successfully guide it at the portfolio level…An entity or offering is in high value where it has significant impact to an organization's environment and where that impact is relevant to the organization's strategy.

(PMI, 2017)

The communication of value to stakeholders should be done via metrics for both tangible and intangible value. As management guru Peter Drucker famously put it, "If you can't measure it, you can't manage it" (Big, 2017). The numbers should tell the portfolio's story of delivering value to the organization.

> But a crucially important point to keep in mind, ... is that it's not the numbers themselves that matter; it's the story the numbers tell… and the message you intend to deliver. Numbers without context are absolutely meaningless and confusing…and, worst case, potentially damaging…but the story, and the meaning, and the humanity, and the emotion behind those numbers. That's what moves people and spurs them to action.
>
> **(Big, 2017)**

Value measurement is not straightforward, particularly for those companies that function in chaotic or highly disrupted environments where they have little to no control. In those circumstances, there may be multiple solutions being pursued simultaneously to achieve the same result knowing fully that they will not all be completed. Some companies may have not seen themselves in this chaotic space until it happens. Here are a few recent examples:

- Many companies had digital projects or even digital transformation as a part of their strategic plans, but then the COVID pandemic forced *many* businesses all over the globe into this mode of operating suddenly – not knowing if business could happen in their retail shops or if employees board a plane to see a client or have clients come to you. Originally thought to last only a few weeks, the pandemic has been ongoing for more than two years. Polling estimates at least 54% of small businesses were negatively impacted with approximately 32.7 million job losses in the first few months (Bartik et al., 2020). Other companies like Amazon who were prepared for digital operations saw profits rise 220% (Weise, 2021). For example, freelancing platform Upwork Inc. saw its biggest year-over-year revenue growth in the fourth quarter of 2020 with gross service volume jumping 33% year-over-year to $727.7 million. At the end of the fiscal year 2020, Microsoft had a market cap of $1.68 trillion, an increase of 39.83% compared to the same period in 2019. Zoom enjoyed a profit of $882.5 million in the last quarter of 2020, an increase of 370% compared to the same period in 2019 (Novicio, 2021).

- The Russian invasion of Ukraine forced business in Ukraine and neighboring countries like Poland, Hungary, and Romania into this chaotic state when war forced approximately 5.6 million or 30% of the Ukrainian population to become refugees crossing over the borders with little or no food or place to go and local businesses looking out their windows at tanks and soldiers instead of potential customers. In Poland alone, about 3 million refugees arrived looking for solace after the invasion. That's equivalent to an 8% gain in the country's population over the course of two months, and it's 45 times the typical annual inflow of migrants (Buehler et al., 2022). In response to this invasion, the iconic American restaurant chain McDonald's closed all of its restaurants in Russia after 32 years (Turak, 2020). The United States and its allies have imposed harsh economic penalties on Russia over its war in Ukraine (Siripurapu, 2022). New York investment bank giant Goldman Sachs is cashing in on the war in Ukraine by selling Russian debt to U.S. hedge funds – and using a legal loophole in sanctions to do it (Allen et al., 2022). "One aspect of the sanctions has received far less attention, even though it can exacerbate the effect of the conflict on civilians. Some of the trade restrictions and financial sanctions pose immediate and concrete challenges to the capacity of humanitarian organizations to work in Ukraine and in neighbouring states." (Gillard, 2022).
- Political upheavals in Sri Lanka yielded electricity shortages, burned down homes and buses and soaring prices of consumer goods in response to the worst economic crisis that country has experienced since the close of WWII which forced a nationwide curfew and a government ordering troops to shoot protestors on sight. The airport was closed effectively shutting down commerce and tourism. "Sri Lanka is the first domino to fall in the face of a global debt crisis" according to The Guardian (Elliott, 2022). Yet the country owes over $50 billion to government creditors such as India, China, and Japan, and private bondholders – and it is no longer making interest payments. "Under the Rajapaksa family's rule, Sri Lanka has incurred a string of Chinese debts, including … white elephant projects that have yielded little to no income. When COVID-19 struck, they ploughed on with sweeping tax cuts as tourism collapsed—wiping out state revenue and personal incomes…Last week, it was down to

its last 24 hours of gasoline stocks. Medicine and food supplies are critically low." (Malloch-Brown, 2022) From a business perspective, Sri Lanka has several regional competitors in the IT sector, including India, Bangladesh, and Vietnam. "We have spent years cultivating relationships with clients, so they are understanding, but there is always a fear that our business will be taken away and given to competitors if we cannot maintain our delivery and quality," says Waidyalankara (Throwfeek, 2022).

These examples show that plans need to be made with contingencies, alternate delivery options, and back up plans. Risk response plans should also be created with external factors in mind – particularly when there are significant risks (and opportunities) due to volatility. Consider that "...companies that enable local response and strategic planning to analyze unexpected occurrences outperform those that use formal ... frameworks" (Copenhagen, 2022).

The Standard for Portfolio Management states,

> To achieve effective Portfolio Value Management, the portfolio manager needs to create a model of the portfolio's requirements, influencing factors and tolerances that will drive the portfolio components towards the realization of the value target...There are two primary feedback loops from the portfolio to the strategy driving it. The first is the one that negotiates the expected portfolio value given the required value resulting from strategy development, assigned budget, enterprise environmental factors, external and internal risks, and organizational risk appetite. The second feedback loop acts more slowly, taking measurements of performance and achieved value and combining those with environmental observation and the organization's purpose and worldview to shape strategy.
>
> **(PMI, 2017)**

The portfolio leaders need to understand where value generation lives and how to bring it forward – not only by component selection but also from seizing opportunities as they are presented.

Finally, the tools and methods used within the portfolio should be adequate for the purpose intended. This doesn't need to be an expensive exercise – especially starting out. It also doesn't need to be intricately complex. Utilize simple templates and processes to standardize, simplify governance, and predict delivery outcomes. Consider what processes are needed and how they apply to varying project types. Some typical processes that support governance are as follows:

- Approval for project to begin
- Change control process for introducing changes to the project
- Approval to proceed at specific milestones or at the end of project phases
- Approval process for receipt and sign off on the final deliverables

Consider the minimum documentation requirements for a project. Project managers can always create more documents as necessary to keep the project moving forward, but there should be mandated documents for every project. Some examples that could fall into your mandatory list include the following:

- Business case
- Project Charter
- Project plan
- Schedule
- Project closure document

There is no "perfect" governance model, methodology, or toolset as the right approach for each portfolio depends on its organization and culture along with the maturity of the team and the culture of project delivery. Select a model that offers just the right amount of structure and support while avoiding the perception of bureaucracy (Ten, 2021).

When in doubt about how to move forward, choose the simpler option. Communicate it well. Train people. Adjust as needed. Review and repeat as necessary.

REFERENCES

Allen, Jonathan; Ruhle, Stephanie; Herman, Charlie. "How Goldman Sachs profits from war in Ukraine, loophole in sanctions." www.Nbcnews.com. Published on 10 March 2022.
Bartik, Alexander; Bertrand, Marianne; Cullen, Zoe, Stanton, Christopher. "The impact of COVID-19 on small business outcomes and expectations." www.PNAS.org. Published on 10 July 2020.

Big, Blue Gumball. "How My Cardiologist Almost Gave Me a Heart Attack (or, the Right and Wrong Ways to Communicate Numbers)." www.bigbluegumball.com. Published on 8 June 2017.

Buehler, Kevin; Chewning, Eric; Govindarajan, Arvind; Greenberg, Ezra; Hirt, Martin; Jain, Ritesh; Mysore, Mihir; Smit, Sven; White, Olivia. "War in Ukraine: Twelve disruptions changing the world." www.mckinsey.com. Published on 9 May 2022.

Copenhagen Business School. "A new approach to enterprise risk management." www.phys.org. Published on 11 Jan 2022.

Ecosys. "10 benefits of project portfolio management." www.ecosys.com. Published on 13 March 2019.

Elliott, Larry. "Sri Lanka is the first domino to fall in the face of a global debt crisis." www.theguardian.com. Published on 9 May 2022.

Gillard, Emanuela-Chiara. "Sanctions must not prevent humanitarian work in Ukraine." www.chathamhouse.org. Published on 30 May 2022.

Great Prairie Group (GPG). "Effective strategic planning: 5 steps to build consensus." www.greatprariegroup.com. Published on June 2019.

Malloch-Brown, Mark. "Sri Lanka is an Omen." www.foreignpolicy.com. Published on 25 May 2022.

Novicio, Trish. "15 companies that benefited the most from the pandemic." www.yahoo.com. Published on 8 March 2021.

Overby, Stephanie. "10 hard truths of change management." www.cio.com. Published on 16 Mar 2022.

Partridge, Matthew John. "The measure of intelligence is the ability to change - Albert Einstein." www.Linkedin.com. Published on 19 October 2020.

Project Management Institute. (2017). *The standard for portfolio management* (4 ed.). Newtown Square, Pennsylvania: Project Management Institute.

Schibi, O. (2013). Why PMOs do not deliver to their potential. Paper presented at PMI® Global Congress 2013—North America, New Orleans, LA. Newtown Square, PA: Project Management Institute.

Sinek, Simon. "Everyone has a Why. Do you know yours?" www.simonsinek.com. Published on 2020.

Siripurapu, Anshu. "Will international sanctions stop Russia in Ukraine?" www.cfr.org. Published on 1 March 2022.

Ten-Six Consulting. "A guide to project governance models." www.tensix.com. Published on 26 April 2021.

Throwfeek, Rehana. "Sri Lanka's economic crisis threatens its dollar-earning IT firms." www.aljazeera.com. Published on 7 June 2022.

Turak, Natasha. "Goodbye, American soft power: McDonald's exiting Russia after 32 years is the end of an era." www.Cnbc.com. Published on 5 May 2020.

Washington, Tim. "PPM 101 – Assess portfolio maturity in order to get there." www.Acuityppm.com. Published on 11 September 2019.

Weise, Karen. "Amazon's profit soars 220 percent as pandemic drives shopping online." www.nytimes.com. Published on 29 April 2021.

4

Establishing a Governance Model for Strategic Portfolio Management

J. LeRoy Ward

The essence of portfolio management is decision-making defined as "the act or process of deciding something especially with a group of people" (Merriam-Webster, 2022). The organization, through its key executives, needs to decide what projects, programs, and other initiatives (hereinafter called projects) to invest in, delay, defer, terminate, change, or modify to meet its strategic objectives. It needs to make these decisions in accordance with a structured and systematic, yet not overly rigid, set of rules and criteria so that they are made rationally and logically, based *primarily* on data and not *exclusively* on "gut feel." The author is not downplaying the value of "gut feel," he is simply trying to keep it in perspective. In fact, "gut feel," commonly called intuition, can indeed be very helpful in decision-making. "Studies show that pairing gut feelings with analytical thinking helps you make better, faster, and more accurate decisions and gives you more confidence in your choices than relying on intellect alone." (Wilding, 2022).

But, it's not just about the act of decision-making alone, it's about making portfolio decisions with due deliberate speed to take advantage of, for example, market-moving news and events, satisfying customers, or meeting critical regulatory requirements. To do so requires a streamlined, customized, and flexible approach, known as Portfolio Governance Management that works with, and not fights against, the culture and best interests of the organization. One useful definition of Portfolio Governance Management is as follows:

> [T]he structure and exercise of authority for the initiatives and the portfolios within the portfolio management domain, which defines

DOI: 10.1201/9781003315902-4

and enables decision-making; assesses metrics on initiatives value and alignment with business strategy; and is responsible for effective and legitimate oversight for the contributions to business success of these portfolios.

(Hanford, 2006, p. 10)

Portfolio Governance Management is based on, and is a manifestation of, a Governance Model (GM), the subject of this chapter.

The Project Management Institute's (PMI) *The Standard for Portfolio Management*, Fourth Edition *(The Standard)*, provides a foundational view of portfolio management in particular by identifying what can be interpreted as a process of portfolio management which, among other things, includes the following six performance domains:

- Portfolio Governance
- Portfolio Capacity and Capability Management
- Portfolio Stakeholder Engagement
- Portfolio Value Management
- Portfolio Risk Management
- Portfolio Strategic Management.

(PMI, 2017, p. 10)

Each of these domains is discussed in detail in *The Standard* and can be very helpful to an organization that is either just beginning to define and implement portfolio management, or one with a semblance of a portfolio management approach but that wants to refine and improve it.

The Standard also includes a nominal portfolio life-cycle example consisting of four stages: Initiation, Planning, Execution, and Optimization (PMI, 2017, p. 23). The GM discussed in this chapter transcends all phases of the portfolio life cycle found in *The Standard*. However, for an organization to "get out of the gate" quickly with strategic portfolio management, the six questions posed in this chapter must be answered initially and cannot wait for the sequential steps as depicted by *The Standard's* life-cycle phases. *The Standard* can be a helpful reference guide in many aspects of portfolio management, but for an organization that is just getting started in portfolio management it does not layout in a very clear way "how" to get started.

Succinctly put, the GM is the "control room" of the portfolio management process. Without it, portfolio management cannot happen in any structured, purposeful, way. Moreover, "a rigorous governance model is critical to help enforce accountability, optimize cross-functional alignment, and escalate issues to the appropriate decision makers" (PWC, 2012, p. 3).

However, *The Standard* places the development of the GM somewhat downstream in its life cycle of portfolio management. In the author's view, this is an inefficient way to begin establishing a portfolio management process. Developing the GM is the absolute first activity an organization should do before anything else in portfolio management, and the reason is simple: the GM outlines and defines all the key elements and activities and scope and direction of the portfolio process. While it is a decision-making model, it is more than that; in short, the GM encompasses every aspect of *how* an organization is going to execute portfolio management. That is why it comes first.

The author acknowledges, and indeed has extensive experience in, managing portfolios where the GM was weak, underdeveloped, or only casually followed. In cases where governance is weak, portfolio management review meetings become events where the loudest voice in the room gets their way, regardless of whether it is directly related to the strategic objectives of the organization, or the boss's projects are the ones that get funded regardless of their real business value. Needless to say, this approach is hardly a professional portfolio management process; rather, it is one that needs to restructure or re-charter itself by implementing a strong, agreed upon, and pragmatic GM.

Therefore, whether the reader is just starting out to develop a portfolio management process in their organization, or is in need of reconstituting or re-chartering portfolio management in a more professional manner, the GM is where we must begin.

Following are, at a minimum, the six key questions that the organization needs to answer in order to develop a pragmatic, sensible, and effective GM.

QUESTION NO. 1: WHAT IS OUR PURPOSE AND WHAT PORTFOLIO WILL WE MANAGE?

This is an important question as it forms the basis for the existence of the portfolio management group in the organization (see the subsequent paragraphs for more on the name of the group). Will it be a decision-making or advisory body? What are the business outcomes we are trying to achieve (i.e., the strategic objectives of the organization)? For example, are we trying to cut costs, increase profits, increase market share, avoid risks, improve quality, or some combination of these and other outcomes? In other words, why are we establishing a portfolio management process, and what do we expect to gain from it?

As way of personal example, more than 25 years ago the author, a former key executive in a large corporate training firm responsible for product strategy and development, was approached by one of his peers with a pressing problem. In the absence of any prioritization and rank ordering of its product-development projects, the organization's Vice President of Product Development literally did not know which projects to work on first. Clearly, given her limited resources she could not work on all projects at once. Yet, each stakeholder held a very strong opinion that his or her project should come first, a common situation in almost every organization.

This situation put enormous pressure on the Vice President and her staff. They realized that while they valiantly strived to perform at a very high level to accomplish what everyone wanted, they knew it was unsustainable. Something had to be done to help the Vice President prioritize the work so that the right type and number of resources could be assigned to execute the projects in the pipeline in a more orderly fashion.

This was exceedingly difficult in an organization where the key executive had one overriding strategic objective: to achieve $100 million in revenue in the upcoming fiscal year. Are financial targets really strategic business objectives? It depends on who you ask. Needless to say, financial targets, in the author's opinion, are only a starting point as regards identifying strategic objectives. What an organization does to reach those targets, generally speaking, is what informs an organization's strategic objectives.

To address the situation, the author recognized that he needed to convene meetings on a regular basis with the firm's key executives in an attempt to prioritize the enormous amount of work in the product pipeline. That was the sole focus of the sessions initially. We did not have a charter, portfolio management plan, performance plan, or the like, nor did we think, erroneously as it turned out, we needed any of these governance documents. We had a serious situation that needed our specific attention and that is what we tackled first. Accordingly, that was our "purpose," and the portfolio to be managed was all of the company's customer-facing (i.e., revenue-generating) projects. It was a good first step in our nascent approach to professionally manage the portfolio. As time passed our "purpose" evolved and expanded to include a number of other business outcomes and objectives.

Once we have defined and agreed to our purpose, we now need to identify the portfolio(s) to be managed. In any organization, or subset thereof, such as a business unit or operating division, it is entirely normal and expected to have more than one portfolio. In fact, the author strongly discourages any organization from including *every* project and program in one portfolio. Greater efficiency, speed, and other benefits accrue from categorizing projects into multiple portfolios to be managed separately.

Consider a consolidated example of a corporate training firm (example represents several training firms) which offers a wide variety of training programs and products to the global marketplace. This firm will have a portfolio of components that include such initiatives as new course development, assessment products, books, other publications, tools, games, and simulations. It may also include projects focused on the different types of training modalities such as on-line, virtual classroom, or video-based learning. The portfolio might also include projects associated with curriculum enhancements. We can see that a wide variety of projects all related to the "product set" of the company can constitute one portfolio.

However, this firm will also have a portfolio of other projects that focus on, for example, internal process improvements, to use a rather broad term. Projects in this portfolio may include the implementation of a new human resources personnel evaluation tool, use of a new Customer Relationship Management (CRM) system, implementing a new email system, and developing a new client-relationship management process to handle global accounts, to name some examples. These improvement projects and initiatives can be conveniently categorized in a second portfolio. The question is should both portfolios be managed as one? The answer is a resounding *no*.

This firm would be wise to have its products managed as one portfolio, and its internal improvement projects managed as a second portfolio. The reason is rather straightforward, and it has to do with focus. Making decisions about its customer-facing product set will, more than likely, include a different set of criteria, mix of resources, and risks than its internal improvement process projects. Also, combining too many disparate projects into one portfolio together can result in difficulties in bringing the right stakeholders together to make decisions and once together can cause meetings to be become too lengthy with certain participants losing interest when projects that they are not involved in or care about are discussed. Having the right group and appropriate number of people at the table is extremely important and will lead to better and more efficient decision-making. One way to accomplish this is to only discuss common projects of mutual interest to all involved. This can be easily done by managing one specific portfolio at a time.

Also assume in the given example that it is a rather small company. As such, the same individuals who are selected to manage the product portfolio might very well be the same individuals who would manage the internal process improvement portfolio. This is perfectly acceptable, and in small companies this is often the case. Yet, the author still advises that the portfolios be managed separately for the reasons so stated.

The author acknowledges that at the highest level in the organization, a particular person, or group, needs to provide oversight for all of its portfolios. At the highest levels, all the portfolios constitute *the* portfolio of the organization. This person or group will, among other important considerations, ensure that the entire collection of portfolio components aligns with the organization's corporate strategy. Figure 4.1 presents a structure of multiple portfolios in an organization, and how they can be managed at various levels. Portfolios One, Two, and Three are managed separately; however, the Chief Executive, through their executive team, or perhaps through an appointed Executive Committee, provides executive-level oversight.

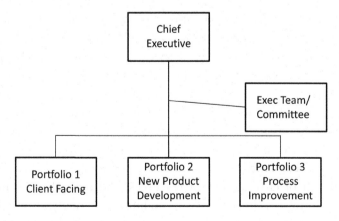

FIGURE 4.1
A suggested approach for managing multiple portfolios.

In summary, the first step in developing a GM is to define the purpose and vision of portfolio management and identify the portfolio(s) to be managed. We can then turn our attention to Question Number 2: Who will manage the portfolio?

QUESTION NO. 2: WHO WILL MANAGE THE PORTFOLIO?

Most organizations form committees or task forces of one type or another to manage the target portfolio(s). They have a variety of names such as the *Portfolio Review Committee, Portfolio Board, Product Portfolio Committee, Portfolio Management Team, Innovation Board*, or a host of other monikers that generally describe the work of the collective team. Regardless of the name of the committee, the key question is "Who should be on the committee?" While it may seem like a simple question, the answer can be quite difficult for many reasons, not the least of which is the political ramifications that such a decision has. From this point forward, the author will use the term "Portfolio Committee" to identify said group or task force.

Best practice in this area suggests that we do not want too many people on the Portfolio Committee because that will slow down decision-making and make the logistics of planning for, convening, and conducting meetings problematic. As anyone who has tried to plan and conduct meetings, we know it is much more difficult to convene a meeting (whether in person or virtual) with 20 people than it is with ten. The author contends it is not twice as hard; it is four times as hard!

At the same time, we do not want too few people on the Portfolio Committee because we need to make sure we have the appropriate perspectives on the issues at hand and have a sufficient number of people who can represent the views of the organization at large. Also, if the team is too small, it will be criticized for serving the interests more of the members themselves than the organization as a whole. But this still begs the question "Who should be on the Portfolio Committee?"

The author's experience strongly suggests that the membership, or at least the "voting membership" (more details on this idea are given in the subsequent paragraphs), be made up of *executives* in the organization, ones responsible for a large operational activity, such as a division or separate business unit, and especially those who manage a profit and loss

(P&L) statement. It is these executives whose performance is typically evaluated by the Chief Executive Officer (CEO), Corporate Board, or majority shareholders and who have a vested interest in the makeup and successful execution of the portfolio.

Second, key executives who support the P&L leaders, and whose organizational units will execute and support the work in the portfolio, might also be included. Buy-in to the portfolio components themselves is very important. These executives will have "skin in the game," and thus are more likely to provide whatever support is required to provide resources to complete the projects that are included in the portfolio in a timely fashion.

Certain organizations include key technical personnel as part of the Portfolio Committee given their deep expertise in the subject matter of the individual projects that make up the portfolio or that are being evaluated for possible inclusion. While the author finds no particular fault with this approach save for the fact that the meetings will have more people than perhaps are necessary. If an organization does include technical personnel, he strongly suggests that such technical personnel do not have a "vote" as to the makeup of the portfolio.

Portfolio management is a business-driven, decision-making process, not a technical one. Technical personnel can sometimes become enamored with technology and lose sight of the business rationale for investing in a certain project. If the reader's Portfolio Committee includes – or is contemplating to include – technical experts, the author suggests that they should not be given a vote as to what the portfolio includes. Thus, a Portfolio Committee can consist of "voting" as well as "non-voting" members depending on the particular composition of the Committee itself (i.e., if technical personnel will be members).

Alternatively, technical analysis of whether a project should be included given its feasibility can be evaluated by such personnel outside of Committee meetings which will not require their attendance. As regards the optimum number of members on the Portfolio Committee, one can refer to *The Scrum Guide* which states: "The Scrum Team is small enough to remain nimble, and large enough to complete significant work within a Sprint, typically 10 or fewer people" (Schwaber & Sutherland, 2020, p. 5). If we consider a Portfolio Committee meeting as a Sprint, keeping the number of members to not more than ten makes sense.

The author once consulted with a client whose Portfolio Committee membership included key P&L executives, product development executives, and technical experts. The Portfolio Manager, who was the Chair of the Committee, complained to the author stating that the technical experts were making decisions based largely on the technical issues at hand, and their personal views on what was "good" for the company's future, rather than reviewing project proposals from a more business-focused perspective. The author recommended that the Portfolio Committee membership remain as constituted but that an "investment subcommittee" be formed whose members should only include the key P&L executives. Additionally, the Portfolio Charter was changed to authorize this investment subcommittee, and only this subcommittee, to make investment decisions. The company enacted the change, and while the technical experts felt somewhat shunned at first, the change had the beneficial effect intended: implementation of a data-driven, not a technical-driven, decision process. We need to remind ourselves that portfolio management is business management, not technical management.

The thorniest issue regarding membership on the Portfolio Committee is whether to include members based on political considerations. In other words, should a particular person be invited to sit on the Committee for "political" reasons"? For example, there may be an individual in the organization who is very well regarded by staff, such as a senior technical expert, whose membership would be held in positive regard by the company given this person's technical expertise. There may be an individual in a business unit who has great influence over the decisions of a key business unit head, a key staffer perhaps. The key staffer's membership may help the Portfolio Committee as a whole because that person would see to it that his or her boss would make decisions quickly. Perhaps the key staffer could be persuaded by the other Committee members to "help" the business unit manager make the "right" decision. There are no "absolutes" in the business of portfolio management, and the author cannot say in any given case that these individuals should or should not be part of the Portfolio Committee other than cautioning the reader that too many members will slow down decision-making.

The one advice the author can provide is that if an individual under consideration for membership has influence or direct authority over the budget, personnel, or physical assets affecting the ability of the organization to execute the portfolio then that person's involvement in the Portfolio Committee is generally warranted. The author can also say without hesitation that we should never include representatives on the Portfolio Committee out of professional courtesy, or to be nice, or to make sure an individual does not feel "left out." The author suggests meeting with the individual to explain the selection criteria, and why he or she was not asked to be involved. In most cases, once explained the individual will accept the rationale and nothing will come of it. After all, the author does not know one person who wants to attend any more meetings than they have to these days!

Heather Colella, Research Vice President at Gartner, suggests answering the following four questions as it relates to "who will manage the portfolio" which can help the reader get started right away.

1. Who will Chair the group?
2. What are the positions of voting members?
3. What are the positions of the nonvoting members?
4. What are the roles of certain specific members (e.g., business process, industry, and technology, legal or financial expertise)?

(Colella, 2014, p. 17)

The author also suggests appointing an "Executive Director" of the Committee if needed. This individual will be responsible for coordinating all the activities of the group including, among other things, documenting the action items and decisions made in each meeting as well as distributing meeting minutes to the appropriate parties. The Executive Director will establish the meeting calendar for the group, remind the participants of each upcoming meeting, collect agenda items (which will be reviewed by the Chair), distribute any documentation for review to each member prior to each meeting, secure appropriate meeting space, and all the other logistical activities of coordinating this important function. Having an Executive Director will allow the Chair, as well as all the Portfolio Committee members, to concentrate on the decisions that need to be made rather than on the logistics and mechanics of organizing and conducting said meetings.

Should the amount of work be too great for the Executive Director to handle comfortably, the author suggests assigning an Administrative Assistant to help. The Executive Director and Administrative Assistant are not full-time positions in most organizations. Such responsibilities can be accommodated by existing staff. The author understands that establishing these various positions takes on the appearance of establishing a bureaucracy at a time when organizations are trying to become more agile. However, the author also knows by experience that if all the

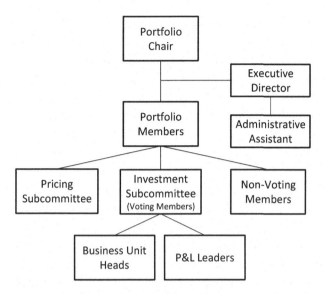

FIGURE 4.2
Example of a Portfolio Committee organizational structure.

"logistics and mechanics" of running an effective portfolio process falls to the Chair, they either won't get done, or get done well.

Figure 4.2 is an example (consolidated from multiple real-life implementations) of a Portfolio Committee organizational chart that addresses a number of the characteristics discussed in the figure.

QUESTION NO. 3: HOW WILL PROJECTS BE SELECTED FOR THE PORTFOLIO?

Now that we have identified the portfolio to be managed, and individuals who will be managing the portfolio, our next decision is to determine the criteria used to decide how any proposed project, program, or initiative (collectively referred to as the projects), will become part of the portfolio. This decision is the primary responsibility of the Portfolio Committee (there are other responsibilities as well). The Committee must also ensure that the process is transparent, agreed to, and followed by all. The criteria to be employed are a core component of the GM.

In fact, a wide variety of techniques and criteria are, and can be, used by organizations to select projects for their portfolio. They fall into two broad categories: quantitative and qualitative. The most common quantitative criteria are financial and include any one of a number of financial analysis models that help the Portfolio Committee decide where to invest the organization's finite financial resources. Such financial techniques include, for example, net present value, break-even analysis, payback period, benefit/cost ratio, internal rate of return (IRR), economic value add, and others that may be unique to the organization. Certain organizations establish a "hurdle rate" that a project must exceed in order to be considered further for execution. Many organizations employ more than one financial model when evaluating a portfolio component.

In addition to quantitative measures, qualitative measures can, and will, be applied as well. Increase in customer satisfaction, employee morale, enhancement of good will, and reputation are but a few such measures. While best practice suggests that all benefits be quantified, there are occasions when that may not be possible. Therefore, selection of portfolio components almost always includes both quantitative and qualitative factors.

If the reader is interested in exploring alternatives for measuring intangibles, the author suggests reading *How to Measure Anything: Finding the Value of Intangibles in Business* by Douglas W. Hubbard (Hubbard, 2007). Among other interesting topics, Hubbard includes seven chapters on measurement methods and techniques.

These quantitative and qualitative factors are an overall part of a business case that is developed and presented to the Portfolio Committee for review and approval. The business case provides the rationale and justification for why the project should be undertaken. But it's more than that. It is also referred to throughout the project's execution to ensure that the benefits initially identified will be delivered based on current performance and in line with financial estimates. A deteriorating business case provides strong justification for terminating any project and removing it from the portfolio.

One other criterion for selecting certain projects has to do with regulatory compliance. For example, organizations in the financial services industry, such as investment banks, hedge-funds, commercial banks, insurance companies, mutual funds, and the like, are governed by a broad array of State and Federal regulations. Many countries, and political entities, such as the European Union, also have strict regulations governing financial compliance. As new laws and regulations governing the operations of financial institutions are written and approved, new projects to satisfy compliance regulations will be added to the portfolio. This is just one industry example.

For example, in the U.S. the Dodd–Frank Act (officially known as the Dodd–Frank Wall Street Reform and Consumer Protection Act), requires financial institutions to operate in specific ways creating a host of regulatory compliance provisions that need to be incorporated into their operations. Such compliance requirements are satisfied through various projects which are included in the portfolio simply because it is mandatory that they be done. If not, the company would be subject to fines and penalties by the U.S. Government. The only decision the company has to make is *how* to do the project not whether it should be done. The project will therefore be included in the portfolio so that it can be carefully monitored by the Portfolio Committee.

The GM not only describes the measures and metrics the organization will use to select any individual project into the portfolio, it will also describe the various methods that will be used to rank the projects in order of importance. This second process is critical to developing the Portfolio Roadmap (PMI, 2017 pp. 35–36). Müller, Martinsuo, and Blomquist in their worldwide study of 133 organizations found, among other things, that successful organizations have a defined process to select *and prioritize* (emphasis added) projects supporting organizational strategy (Müller et al., 2008).

There is one other point regarding the specific projects that will make up a portfolio. The author has observed that even when an organization has a mature portfolio management process, not *every* project in the organization is, or should be, formally included in the portfolio for professional management purposes. There may be projects of such low monetary value that specific selection rules do not apply, yet they are needed as part of a larger effort. To be sure, this is completely at the discretion of the organization. The author consulted with one organization where projects valued at less than $200,000 were not included in the portfolio for oversight and monitoring purposes at the Portfolio Committee level. These projects fell under a different and separate management regimen monitored through other means. Of course, such projects are part of the organization's portfolio, but they do not necessarily have to be part of a formal portfolio process managed by a Portfolio Committee.

The reader can readily see that the rules of the road for defining the universe of projects that will be part of a portfolio can be as varied and unique as the organization itself. This is an important aspect of defining a GM, and the organization's leaders need to decide just which projects will become part of a portfolio to be managed in a structured, systematic way. These decisions are not easy, but they must be made at the outset to get the portfolio process under way. The organization must also recognize it can change the rules at any time as well if, for whatever reason, the selection and prioritization process is not efficient or is not achieving its intended results.

QUESTION NO. 4: HOW WILL WE MAKE DECISIONS AND RESOLVE CONFLICTS AS A PORTFOLIO COMMITTEE?

As stated earlier, the essence of portfolio management is decision-making. What projects should be included or not? What projects should be delayed or deferred? What proposals for new portfolio components should an organization entertain? And the list goes on and on.

The author's experience in establishing and coordinating a Portfolio Committee for more than 20 years clearly shows there is no shortage of decisions, large and small, of critical importance or completely mundane, that needs to be made by the Portfolio Committee. In order to make decisions quickly, there has to be clear rules that everyone agrees to and follows. One thing the Portfolio Committee should never do is to subjugate their decision-making to a financial model or other analytical or financial decision-making technique. Models and techniques do not make decisions, people do. Models and techniques though are helpful for the Portfolio Committee as a way to start a conversation about the decision to be made.

Even if the Portfolio Committee is presented with clear and compelling financial justification (e.g., has an IRR of >25%) indicating that a particular project should be included in the portfolio, a "slam dunk" if you will, certain Portfolio Committee members will disagree and will argue that other projects be included instead. Many discussions at Portfolio Committee meetings are not necessarily about proceeding ahead with one project or terminating another. They are more about the broad strategy and direction of the organization and how the collective projects are meeting those goals. Regardless of the nature of the conversations or disputes, the Portfolio Committee needs to have a way to resolve these conflicts. So, how is this accomplished?

As stated earlier, there are no "absolutes" in the business of portfolio management. The author can no more recommend a specific way to resolve such conflicts other than to suggest that an escalation process needs to exist to resolve disputes, otherwise a stalemate will occur and time will be lost. The rules established must be palatable to all, and in fact, should be developed by everyone on the Portfolio Committee representing its broad consensus. Establishing rules and forcing them on the members will only create antagonistic feelings among the group. Members will either aggressively, or worse, passively, resist the process causing even greater problems and delays in the future. How an organization establishes such rules can oftentimes be more important than what those rules actually are.

Some Portfolio Committees operate on a very simple model: majority rules. This is an easy and quick way to make decisions, but the organization needs to have the discipline (i.e., intestinal fortitude) to abide by the voting outcome. Too often, these votes are nothing more than "soft decisions," which are then reviewed at the next Portfolio Committee meeting and possibly overturned. Other Portfolio Committees will cast an informal vote that will then be reviewed by a more senior executive or a Portfolio Oversight Committee, one level above the Portfolio Committee. The final decision will rest with either of these two. As such, the Portfolio Committee is then acting in an advisory capacity to a more authoritative entity rather than a decision-making body (see Question Number 1).

The author has had experience with the latter. One client, a major pharmaceutical corporation, had a process where each member of its Innovation Management Board (its version of a Portfolio Committee) voted on which projects to initiate or kill, but in the end, all major product decisions, those that involved a substantial amount of investment, or would otherwise redirect the product strategy of the organization, needed to be reviewed and approved by the CEO (who did not sit on the Portfolio Committee). Of course, this process caused delays, but it was an acceptable price to pay to make sure the CEO agreed with, and funded, the proposed project.

The reader's GM must have a way to break logjams and stalemates among Committee members. The process for doing so must be discussed and then put into practice by consensus of the Portfolio Committee and other key executives. If no method exists to address these difficult situations, then individual members of the Portfolio Committee will lobby key stakeholders to advance their agenda. This lobbying effort can result in hard feelings by promoting a zero-sum, winner take all, approach to decision-making. Moreover, while the result might be in the best interest of certain stakeholders, it may not be in the best interest of the organization as a whole. We should avoid the practice of "horse trading" because this practice promotes the interests of individual members of the Committee and certain stakeholders quite possibly at the expense of the needs of the organization.

The author believes that a portfolio management process should not necessarily be an entirely democratic one. At some point, an individual with a strong personality, and the will to exercise it, combined with a senior title, will need to step in and make the final decision. That person will be either a senior executive or an oversight committee that essentially serves as a proxy for a senior executive. However manner in which decisions are made, it is important that the approach be clear and transparent to all.

QUESTION NO. 5: HOW OFTEN WILL WE MEET AND WHAT RULES WILL GOVERN OUR MEETINGS?

Strategic portfolio management is an active and continuous management process. Accordingly, the Portfolio Committee should meet on a regular basis to discuss matters of importance including:

1. Reviewing projects proposed to be included in the portfolio
2. Reviewing and evaluating ongoing projects to decide if they should be continued, delayed, or terminated
3. Assessing the current order of priority to see if it should change
4. Discussing resource issues affecting the scheduled completion of the various portfolio components
5. Monitoring the identified portfolio risks to determine if they are likely to occur or to eliminate ones that no longer pose a threat
6. Scanning for opportunities for greater efficiency and effectiveness in portfolio performance
7. Assessing the quality of the portfolio management process used by the Portfolio Committee
8. And, any other matters of particular and unique interest to the Portfolio Committee as it relates to the portfolio and the portfolio management process

The key question is how frequent is frequent enough? The author suggests that the Portfolio Committee meet no less than once per quarter in any given calendar (or fiscal) year. To be sure, in an ever-changing business environment, much can happen in three months, and we need to keep our collective fingers on the pulse of the organization's portfolio activities. In certain business environments, meeting monthly is advised. In other, fast-moving business environments, where the "need for speed" is paramount, even monthly is not frequent enough.

The frequency of the meetings will be dictated by the portfolio components themselves. If, for example, the organization's projects tend to run three months or less, then monthly, or more frequent, meetings should be held. If they are longer, then less frequent meetings will suffice. In no case though, should the Portfolio Committee meet less than once per quarter.

The author suggests convening meetings on the same day, time, and for the same time period each time a meeting is held. For example, the Portfolio Committee will meet on the second Thursday of every month for two hours from 9:00 AM – 11:00 AM. This approach allows the Executive Director to place this meeting on everyone's calendar, and it is firmly established for each Committee member for the year.

Meeting attendance should be made mandatory, and only under extraordinary circumstances should a voting member miss a meeting. Having key executives unable to attend slows down the decision-making process. Time is of the essence in business today, and decisions need to be made quickly to capture market share or beat a competitor to market. Oftentimes an executive will send a "key staffer" in his or her place. The author's experience shows that this does little to speed up decision-making. At best, these individuals are "note takers" and "seat fillers" for the executive. This does little good to the Portfolio Committee which is charged with making key, and difficult, decisions in a timely manner.

The author suggests that if a key executive is unable to attend a Portfolio Committee meeting, then he or she should inform the Portfolio Chair well in advance (especially if meetings are held only once per quarter) so that they can meet separately to discuss the issues that will be raised at the meeting and express his or her opinion on key decisions that will be made. Of course, should a rather significant number of voting members not be able to attend any specific meeting, it should be re-scheduled when a quorum will be in attendance. A "quorum" should be defined in the meeting rules.

The meeting agenda should be standardized for efficiency. The author suggests starting each meeting with a review of the minutes of the last meeting and the status of any action items assigned. There are two broad categories of agenda items that will follow the review of the meeting minutes: one is a review of the status of ongoing portfolio components, and the other is a review of proposed projects to be included in the portfolio. For the most part, the order in which these are discussed is immaterial. Some organizations prefer to discuss on-going projects first, while others opt to review proposed projects.

Regarding the review of ongoing project status, we must be careful to discuss only projects that, for whatever reason, are of interest or concern to the Portfolio Committee. From a pragmatic perspective, the Committee will probably not have sufficient time to discuss all projects' status, and, it not necessary to do so. To be should review projects that are on a "watch" list, are in danger of not meeting their goals and objectives, whose priority might have changed because of certain market conditions, or, whose variances to time, cost, and scope have exceeded acceptable tolerances.

Additionally, to accelerate decision-making, the Portfolio Committee members should receive project dashboard information on all projects prior to the meeting, and, they should read the reports! If this important information is provided in advance, less time will be spent in the meeting discussing the information itself, allowing more time to discuss how to address the issues at hand.

With respect to the discussion on proposed projects, prior information provided to the Portfolio Committee is, again, helpful, but if only the members read it. Organizations would be wise to have specific rules on what data and information the project sponsor needs to provide the Portfolio Committee in order to make its decision. The author suggests inviting the sponsors of the proposed projects, if different from the executives on the Committee, to "pitch" their ideas to the Portfolio Committee. This will enable the participants to have a full discussion on the merits of the proposed project.

If a proposed project supports the organization's strategic objectives and satisfies the merits of the business case, the Portfolio Committee can decide at the meeting to include it in the portfolio. However, in many instances, the Committee will need time to think through the proposal, and how any specific project might affect the priorities of other portfolio projects. Even if a project appears on its face to be beneficial to the organization, its inclusion might require a re-prioritization of a number of portfolio components that might be very disruptive.

Portfolio management is not as easy as some would make it out to be because it is not about the management of a collection of individual projects; its focus is on assembling that collection of projects whose total combined output provides the greatest benefit to the organization as a whole. The main guidance as to what to include or not include in any portfolio, above all other guidance, is the stated strategic goals of the organization. This is the first test to be applied to any project.

DECISION NO. 6: WHAT WILL BE THE ROLE OF THE PMO IN PORTFOLIO MANAGEMENT?

Having worked in the project management profession for more than 40 years, the author has observed many different implementations of Project Management Offices (PMOs) globally and in all industry sectors. Clearly, there is no "standard" PMO, notwithstanding the fact that many PMOs provide the same types of services to the organizations they serve. There are certain PMOs, which are, in fact, actively involved in portfolio management. These tend to be PMOs that have been in existence for a number of years and ones that have direct management oversight of the project and program managers in the organization. In addition to being responsible for the execution of the projects and programs in the portfolio, their involvement in portfolio management tends to be with providing the Portfolio Committee with the important status information regarding the various portfolio components. Typically, a PMO head is not a voting member of the Portfolio Committee, but is its Executive Director or Advisor. Additionally, the PMO may be called upon to provide assessment services helping the Portfolio Committee analyze a project proposal to determine if it passes muster.

Other PMOs have had little if any involvement with portfolio management in their respective organizations. These tend to be very small PMOs of only two or three people and have no operational responsibility for project execution. They exist primarily to promote the standardization and application of project management practices, software, and other tools, and to coordinate training activities.

There also exists somewhat of a hybrid type of PMO as illustrated by a former client with whom the author consulted for more than ten years. The client, a global information technology organization, had an EPMO (Enterprise Program Management Office) headed by a Vice President who reported to an executive vice president who was the P&L leader of one of the organization's major business units. The EPMO employed approximately six professionals focused on several key areas of project management: training, methodology, tools, and the "health of the portfolio." It was this last responsibility that brought it squarely into the activities of portfolio management of more than 2,000 active projects generating multiple billions of dollars in revenue worldwide.

As regards the "health of the portfolio" this EPMO was a participant in helping to select various projects for the portfolio and then providing data and information on the status of these ongoing projects to upper management through a sophisticated dash boarding process. The EPMO would continuously assess the health of the portfolio's various key projects by conducting audits and risk reviews and reporting its findings to the business unit executive team.

For many years, the EPMO was seen as performing a critical role in the organization's portfolio management process. In this example, we have a PMO acting as a "staff" function operating at a very high level in the organization working side-by-side with key business unit executives to manage its vast portfolio of information technology service projects.

If the reader is developing or improving a portfolio management process, the author advises them to review the operations of any existing PMO within their organization to determine if its involvement would be beneficial to the overall implementation and application of portfolio management. In many cases, a decision regarding including a PMO or not in portfolio management is based on the professional experience and business acumen of the person or persons working in the PMO, rather than on the fact that there is a PMO.

DISCIPLINE AND ENCOURAGEMENT: ESSENTIAL INGREDIENTS IN IMPLEMENTING A GOVERNANCE MODEL

Portfolio management is one of the most important critical functions of any executive team. When done well, it results in the selection and execution of the optimum collection of projects, programs, and initiatives designed to meet the organization's strategic goals and objectives. And by best, the author means that collection of work that offers the maximum return that conforms to the organization's risk profile.

It is not easy to establish a portfolio management process for the simple reason that everyone seems to have their own ideas as to how it should work, what criteria should be used, who should be involved, what portfolios should be managed, how decisions should be made, and all the other issues that go along working with a group of peers to achieve a common vision. But simply because it is challenging does not mean we should not try to do it, and do it well.

Establishing rules is one thing, abiding by them is another. Organizations can tackle the hard part of the process by answering the six questions presented in this chapter. But the harder part of the process is the diligence and discipline required to make those rules come to life and work for the benefit of the organization. Executive "bad behavior" is the root cause of failure of any portfolio management process. Executives can sometimes place themselves above their very own rules because of their sense of self-importance, omnipotence, impatience, or just an "I know better" philosophy.

Time and time again the author has witnessed executives skirting the rules about providing business cases for the projects they want to do, initiating projects without due authorization, re-prioritizing work based on their own agendas, and only paying lip service to the portfolio management process. This is organizational self-defeating behavior because it affects everyone involved in project execution.

The only way for portfolio management to succeed in an organization is for those key individuals who will be responsible for portfolio management to be actively involved in developing what the process is going to be. And, they then need to monitor themselves to make sure they are all abiding by their collective agreement. One way to make the process much more palatable is to introduce "just enough" portfolio management to get started. Do not introduce a very heavy, documentation driven, portfolio process into an organization that has been used to making decisions on the fly with just a few people and no records! Everyone around the table needs to be convinced that professional portfolio management is in their best collective interest.

REFERENCES

Colella, H. (2014). Top recommendations to design an effective governance process. *Gartner Webinar*. March 27.

Hanford, M. F. (2006). Establishing portfolio management governance: key components. *IBM Corporation White Paper*. Retrieved on March 3, 2014, from http://www.ibm.com/developerworks/rational/library/oct06/hanford/

Hubbard, D. W. (2007). *How to measure anything: Finding the value of intangibles in business*. Hoboken, NJ: John Wiley & Sons.

Merriam-Webster. (2022). Merriam-Webster online dictionary. Retrieved on June 8, 2022, from https://www.merriam-webster.com/dictionary/decision-making

Müller, R., Martinsuo, M., and Blomquist, T. (2008). Project portfolio control and portfolio management performance in different contexts. *Project Management Journal*, 39(3), 28–42.

Project Management Institute (PMI). (2017). *The standard for portfolio management*. 4th edition. Newtown Square, PA: Project Management Institute.

PricewaterhouseCoopers. (2012). How governance and financial discipline can improve portfolio performance. Retrieved on March 10, 2014, from http://www.pwc.com/us/en/increasing-it-effectiveness/publications/strategic-portfolio-management.jhtml

Schwaber, K. and Sutherland, J. (2020) The scrum guide: The definitive guide to scrum: The rules of the game. November 2020. Retrieved on June 14, 2022 from https://www.scrum.org/resources/scrum-guide?gclid=CjwKCAjw46CVBhB1EiwAgy6M4k0kV-PH_UYa-8Y7WlKqzZ9SFkCZuJPtHMWTjl5g4Rs7dIM9tRpzuRoCouYQAvD_BwE

Wilding, M. (2022). How to stop overthinking and start trusting your gut. *Harvard Business Review*. March 10, 2022. Retrieved on June 10, 2022 from https://hbr.org/2022/03/how-to-stop-overthinking-and-start-trusting-your-gut

5

Portfolio Governance

Jennifer Young Baker

Portfolio governance involves the processes used to identify, select, prioritize, and monitor projects within a company or line of business. The foundation of these processes should be the strategy and guiding principles of
the company. With a firm foundation, ongoing governance and oversight
are able to proceed down the path to arrive at the strategic destination.
Portfolio governance is a critical capability in the best of times. Portfolio
managers must navigate their projects toward achieving the "…overarch-
ing strategic objectives" (James, 4). A well-structured governance model
provides the following:

- Accountability
  - Clear roles, responsibilities, and accountabilities
- Transparency
  - Clarity of stakeholders and financial authorities
  - Clear scope
  - Regular meeting cadence (schedules)
- Integrity
  - Ethics
- Protection
  - Dispute- and conflict-resolution escalation channels
  - Empowerment of individuals to "do the right thing"
- Compliance
  - Clear procurement processes
  - Clear adherence to all regulatory, legal, and policy requirements
- Availability
  - Clear reporting and information flow

DOI: 10.1201/9781003315902-5

- Flexibility
  - Ability to adapt to changing business and organizational needs
  - Capability to accommodate shifts in organizational size and complexity
- Retention
  - Performance and benefits
  - Obvious delivery model overlays
- Disposition
  - Clear transition from project/program to operations

The intention of portfolio governance is to facilitate strategic decision-making that will enable delivery of strategic goals and objectives. The basis for those decisions should be consistent and transparent with risk and opportunity taken into account. There are a few ways to accomplish this set of outcomes with defined processes to support it. For many organizations, governance at the portfolio level is an opportunity to ensure corporate and regulatory compliance is adhered to by the portfolio components. Additionally, there are two sides to portfolio governance – a component side and an internal side.

The component side deals with how each portfolio component (projects, programs, sub-portfolios, and operational work) is managed and supported throughout its lifecycle. The internal side is how the portfolio manages itself as well as the processes and framework that define the portfolio and its composition. We will start the discussion with the components and the role governance plays in component delivery.

First, all decisions require information and that information must be accurate. While the people closest to the information may not be able to make those decisions, they will need to see that the information is presented to those who are making the decisions and, in many cases, make a recommendation or provide some level of analysis. It also must be clear which decisions the component teams are allowed to make, and which need to be escalated and to whom. A decision matrix is an excellent way to clearly align the teams about what decisions they can make as well as let the leadership know which decisions will be escalated for resolution.

The concept of a decision matrix is to list out criteria and options with prioritization. It should clearly show which decisions the team and leaders are and are not allowed to make. The matrix should also show the decision-making path for escalation. The sample matrix provided here is for a program component within the portfolio. It is important to keep in mind that some decisions may have boundaries established by corporate policy such as signature authority related to corporate expenditures. These rules help explain why a leader may decide in one circumstance but escalate in another (Table 5.1).

The decision-making process within portfolio components should be closely monitored by the individual sponsors and the portfolio management team. Each sponsor is responsible for the successful delivery of the individual components. The portfolio leadership has accountability to ensure that project sponsors are trained and know what their responsibilities and duties are within that role. According to the Wellingtone's "The State of Project Management Report 2019", the project management challenges with the most significant increases compared to 2018 show that lack of governance increased 26% and poorly trained sponsors increased by 33% (Wellingtone, 5).

Project Management Offices (PMOs) that excel with the definition and execution of strategy have the ability to link the strategy of corporate entities to those of the larger corporate strategic plan. They articulate that strategy in a way that is not only operationally relevant but also is consistent with corporate culture. This allows portfolio management to make appropriate decisions regarding assets within their portfolio. The culture of the organization "...must reinforce all aspects of the project portfolio governance methodology" (James, 10).

It is no surprise that other institutions have noted similar survey and report outcomes. According to PMI's "2018 Pulse of the Profession" report, the number one driver of project success is investing in actively engaged executive sponsors. They went on to note that inadequate sponsor support is the primary cause of project failure for a bit more than one in four organizations (Success, 6). This clearly shows that sponsorship training is important and has a direct link to the successful outcome of the component's work.

Among the duties of the project sponsor are administering or approving project finances and changes which involve making key decisions and/or recommendations, facilitating stakeholder engagement, and supporting the communication processes. Many sponsors sit on one or more governance boards. These boards can include steering committees, change control boards, or be part of the operational management structure within the organization. Having a governance board in lieu of a single decision-maker can be a way by which an organization can instill a sense of checks and balances within the decision-making process as well as an avenue for communication of upcoming changes.

TABLE 5.1

Decision Matrix

	Project Manager	Program Leadership	Sponsor or Steering Committee	Portfolio Leadership	Senior Leadership
What can project teams decide?					
Decisions that align with current, approved scope	Decide	Endorse	N/A	N/A	N/A
Decisions that align with corporate processes that do not impact items listed in the table	Decide	Endorse	N/A	N/A	N/A
What must project teams escalate?					
Decisions that impact other projects or initiatives within the Program	Recommend	Recommend or Decide	Endorse or Decide	N/A	N/A
Decisions that impact other components in the Portfolio	Recommend	Recommend	Recommend or Decide	Endorse or Decide	N/A
Decisions that impact schedule such as major milestones or critical path (Positive or Negative)	Recommend	Recommend or Decide	Recommend or Decide	Recommend or Decide	Decide
Decisions that impact scope other than what is approved	Recommend	Recommend	Recommend or Decide	Recommend or Decide	Decide
Decisions that impact safety, regulatory, or compliance	Recommend	Recommend	Recommend or Decide	Recommend or Decide	Decide
Decisions that impact the business beyond the approved Program Management plan	Recommend	Recommend	Recommend or Decide	Recommend or Decide	Decide
Decisions that impact budget	Recommend	Recommend	Recommend or Decide	Recommend or Decide	Decide

The function of each of these boards varies from organization to organization. Generally speaking, a Steering Committee's role is to provide advice and ensure successful delivery of the project and its expected outcomes. Normally, the steering committee is not close enough to the project activities to provide specific input or guidance without significant information.

Typically, the sponsor is a member of the steering committee. The collective group empowers both the sponsor to make decisions and the project manager to execute the project successfully.

In many circumstances, there is a separate change control board. While this may seem excessive, for some organizations it may be a necessity for regulatory compliance purposes. Though moving a project implementation date seems like a minor change, its downstream impacts can be significant. Particularly with technology implementations and the integration of software, release management is a crucial component to any change. Even when the software and systems seem independent, the combined impact of the collective changes may seem excessive to stakeholders and create what is commonly called change collision in the change management space.

For an example of the need to have release management embedded within the portfolio governance process, the author experienced first-hand the epitome of change collision and the havoc it can create. One bank acquired another. The acquiring bank operated in a siloed fashion and scheduled changes believed to be mutually exclusive. However, within the acquired institution not only the systems were interconnected but also for the staff there was the added component of acquisition. Couple this with changes to how activities, processes, and operations were embedded into the new combined environment and the change in culture, and the recipe for change collision was complete. This circumstance led to payroll systems, information security systems, and distributed computing systems – all being changed on a single day, i.e. December 31st. This day was selected because it was not only the date on a weekend but there was an extra day if needed to correct any issues with the implementations. Because the changes were made after hours and the next day was a bank holiday, it took two days to feel the full impact of what had been done. Thousands of the acquired bank's employees, including the author, did not get paid at year end and could not log into any system when they returned to work on January 2nd. Help-desk employees could not provide assistance because they were also unable to log in, yet the queues were flooded with calls.

It took weeks to sort everything out, serving a massive blow to employee morale.

Another key component of portfolio governance is the concept of phase gating or stage gating. There are many adaptations and variations for a myriad of reasons – product management and project delivery methodologies for example. The intent for these gates is to ensure that the project and programs are ready to move to the next phase or stage in their lifecycle. For some lower maturity organizations, this is a check the box type exercise, but the intent is far from it. The goal is to ensure that appropriate planning has been done and decisions have been made to ensure successful delivery. The gates should also serve as an additional escalation point and avenue for communication. Additionally, where there are compliance requirements at stake, stage gates present an opportunity to ensure that the corporate policies and government regulations are being followed. Also consider that once a project gets to the execution stage, the cost influence factor begins to drop significantly, and the project expenditures rise conversely. Instilling checkpoints at key decision points within the project lifecycle allows for cost and impact control for the organization.

The documentation reviewed at a phase gate or stage gate review will vary by phase of the lifecycle, regulatory requirements, and any changes which have incurred or are requested to move forward (see Table 5.2). Gate meetings occur at the phase transitions and are scheduled by the rate of progress and complexity of the project. These gates can have a variety of outcomes which can include any and all of the following: execute the next phase per the plan, rework one or more of the last phase's deliverables, or to revise and resubmit the plans for the phase. Another outcome of a gate review can be to fully stop/end work or further progression if the purpose and benefit are no longer consistent with the risk, required resources, or enterprise strategy.

The rules around governance should lay out the framework for ownership of projects and portfolios, defining the steering committee structures and roles and assigning responsibilities. Clarity about governance will be a critical first step to addressing decision rights and will provide a foundation for further improvements. "The policies to drive strategy should be concise and unambiguous" (James, 6). They should set expectations at each stage of the project lifecycle, effectively removing the element of surprise from project reviews for any project manager. The policies should contain a rule set which defines where items are non-negotiable and where latitude is

TABLE 5.2

Governance Expectations by Project Lifecycle

	Initiate	Planning	Execute	Commission	Closing
Question for Gate Review	Does the idea merit move forward?	Does the plan justify extensive investigation and execution investment?	Is the product, service, or asset ready to be commissioned?	Is the product, service, or asset ready to transition to normal operations?	How did the project perform based on projections?
Documentation Requirements	Documentation should include the initial project charter and high-level business case. Preliminary market or engineering assessments may also be required.	Documentation should include the detailed business case, quantifiable risks, milestone schedule, and value proposition. It may also include technical feasibility assessments and prototype requests.	Documentation may include a detailed implementation plan, change management plan, test plans, risk contingency, and open issue log.	Documentation may include validation steps and turnover and transition documentation.	Documentation may include signed and accepted deliverables, lessons learned, and financial closure documentation.
Concept	Quick, inexpensive preliminary investigation and scoping of the project – desk research. The need for prototype development should have preliminary requirements.	Detailed investigation involving primary research – both market and technical – leading to a business case, including product and project definition, project justification, and the proposed plan for development.	The actual detailed design and development of the new product or service, including the design and testing of the operations or production process required for eventual full-scale production.	Commercialization – beginning of full-scale operations or production, marketing, use, or selling of product or service.	Assessment of project or program execution performance, including closure of finances and contracts.

provided. These rules balance the need for local customization with guidelines that ensure adherence.

The governance framework should consist of definitions which guide strategic alignment, financial adherence, risk appetite, and resource requirements. Project scope should clearly articulate how investments should support the top business unit priorities. Financial direction should provide clear objectives in terms of investment performance (Center, 4). Stakeholder risk appetite should govern investment and management decisions. These principles should also provide clear expectations for maximizing capacity given scarce resources, including financial, human capital, infrastructure, and partner resources.

Strategic shifts or changes need to be reflected within the governance model. The roadmap will be adjusted, which can have substantial downstream impact and trigger the need for reprioritization. According to Project Management Institute (PMI) survey respondents, change in organizational priorities was the most frequently cited reason for project failure (PMI, 3). This research was echoed by PM Solutions showing PMOs were most successful had been able to consistently align organizational strategy (Center, 2). An example of shifting strategy could come from internal sources of change such as a merger announcement, a senior leadership change, or external sources of change such as a significant legal or regulatory shift or a change in a political party during elections. Portfolio leadership must be able to accommodate for changes and shifting priorities in order to remain successful. "Guiding principles, combined with a comprehensive strategy, effectively shape all other elements of project portfolio governance. The decision-making frameworks, monitoring processes and, ultimately, culture all flow from these principles" (James, 6).

Some type of portfolio-level governance committee should oversee the gate meetings. However, in some cases a program may conduct stage gates for its components and then undergo stage gating for the program's work. The committee members for these sessions are senior managers, sponsors, stakeholders, and other project managers or regulatory representatives. Because decisions or general discussions may include resources, it is customary for the project team to not be included in these sessions. Once the decisions are made, the project or program manager will inform the team of those decisions.

Another consideration in the area of phase gates is to utilize tiered governance which means to have varying requirements based on the expenditure as well as the level of risk, impact, and complexity. The intent is to provide greater scrutiny on the most complex changes while allowing routine changes to move forward as planned. For example, a project that is completing a series of upgrades on server operating systems is something that is done sometimes as often as annually and offers benefits (though limited), but it can be also considered a routine work. This type of project is minimal risk for most organizations and can function with minimal supervision and checkpoints as well as a more junior project manager due to the repetitive nature and simplicity of the project work to be executed.

This concept of tiered governance is also a way that a portfolio can ensure that its most seasoned project management staff can be assigned to the most complex efforts and allow more junior staff members to effectively manage project efforts with lower complexity and risk. One way to complete this type of effort is to establish a matrix for both project/program cost and complexity paired with experience and education of staff. This example is modeled after a risk heat map (Table 5.3).

An example of the most complex and costly initiative could be implementing a new ERP system which is a notable change to the operations and culture of an organization. It is very risky and carries inherent challenges. This is a circumstance where standard phase/stage gates may not be enough. Additional compliance checks or audits may need to be added as well as organizational change management support to ensure that activities are professionally managed. This type of implementation is notorious for wreaking havoc. CIO Magazine recently published an article entitled "16 famous ERP disasters, dustups and disappointments" discussing the string of failed implementations in large and mid-sized companies from multiple industries, including Revlon, MillerCoors, National Grid,

TABLE 5.3

Career Path Matrix by Project Complexity and Cost

	Low Complexity	Moderate Complexity	High Complexity	Extreme Complexity
Highest Cost Band	Level 4: Mid-level	Level 5: Upper Mid-level	Level 6: Senior	Level 7 (Highest): Most Senior
Large Cost Band	Level 3: Lower Mid-level	Level 4: Mid-level	Level 5: Upper Mid-level	Level 6: Senior
Medium Cost Band	Level 2: Junior	Level 3: Lower Mid-level	Level 4: Mid-level	Level 5: Upper Mid-level
Lowest Cost Band	Level 1 (Lowest): Most Junior	Level 2: Junior	Level 3: Lower Mid-level	Level 4: Mid-level

Target, and Hershey's. The very first example in this article discusses controls and governance problems identified prior to implementation which contributed to the failure (Fruhlinger et al., 2020). Other examples in this category include megaprojects and bleeding edge project work which brings forward the next example.

There are some projects that are challenged from inception and poor governance merely exacerbates the problem. One such example is the Duke Energy Edwardsport IGCC Project in Edwardsport Indiana that kicked off in 2007. According to the Indiana Office of the Utility Consumer Counselor,

> The 618-megawatt integrated gasification combined cycle (IGCC) facility (was) designed to convert coal into a synthetic gas. Remaining nitrogen oxide (NOx), sulfur dioxide (SO2), and mercury emissions are removed from the gas before it is burned to generate electricity. The new facility replaced a 160-megawatt station at Edwardsport that had been in operation since the 1940s.

This megaproject was a type of plant conversion that had never been done before. The desire to be first in industry to accomplish this feat blinded internal stakeholders to the inherent risk with what is sometimes called being on "the bleeding edge of technology". The original estimated cost in 2006 was $1.985B US. By 2008, it had ballooned to $2.35B US and again in 2009 changed its cost estimate to $2.88B US. (Undisclosed internal estimates were much higher.) In addition to the massive cost overruns, the project was delivered significantly late – not being commissioned for service until 2013. Duke Energy reported that they had just recovered its project costs in 2020. As soon as 2010, multiple lawsuits as well as criminal charges were levied against the organization as well as responsible individuals (Duke, 2022). The project management staff assigned to this effort were ill-equipped to handle the vast complexities assigned with this work. The outcomes of this litigation would have significant repercussions throughout the organization. One of the mandated changes included establishing a Project Management Center of Excellence (PMCOE) in 2013 with the implementation of tiered governance for projects. The PMCOE established multiple standards from the most complex to the simplest efforts that included compliance audits throughout the project lifecycle to prevent some of the challenges that were evident here as the project was underway. A career path framework was also established to allow the most senior staff to manage the most complex efforts. The assigned career path role was tied to project cost and complexity and reinforced with governance efforts throughout the company.

This example should make one wonder how this "first of a kind" project was selected in the first place given that utilities are known as risk-averse organizations. Project selection within an organization is a portfolio function that directly tied to the purpose and goals of the portfolio – the reason it exists. Starting without a goal and a plan is a recipe for failure. Just like with a project, a portfolio should begin with a charter and a plan. Project selection has its own lifecycle – 1. Select the Right components, 2. Do the selected components Right and 3. Implement the Right systems and tools to monitor and adjust. An example demonstrating this lifecycle can be seen in many regulated companies that complete projects in order to meet new or changing regulatory and/or legal requirements. By doing this, portfolio managers have selected the right projects to avoid regulatory fines and reputational risk. This is done most of the time as a project to ensure that the proper attention is paid to the work being completed. It also provides auditors the opportunity to review what was done and how it was done by reviewing project artifacts. Those artifacts should embody the standards for how the projects should be executed to provide value to the organization. This is executing projects the right way. The systems or "checks and balances" that are implemented to ensure compliance for the regulatory changes would be the last arc in the circle. The core of this philosophy should be embedded within the culture of the organization.

The portfolio charter should discuss the strategic goals and objectives that a portfolio aspires to accomplish and what benefits that it will bring to the organization. The sponsor of the portfolio as well as its leadership and stakeholders should all be discussed within the document and approved to authorize the portfolio. An initial budget should also be included. Just as with the sponsor of a component, the portfolio sponsor has a key role in escalating risk, opportunity, and challenges that the portfolio may face.

A portfolio management plan should be created showing the intent to deliver benefits and the path to get there. The plan should include at least the types of components that are expected, if not the specific component names that will deliver the goals. The goals should be derived from the organization's strategic plan or directly support its delivery by demonstrating how the portfolio will facilitate collaboration, risk management, and meeting the strategic objectives. The goals should be measurable and documented with key performance indicators (KPIs) with an expected timeline. By clarifying KPIs, priorities, and major strategic concerns, portfolio management can focus on the extremely specific needs and goals of the organization. Other essential parts of this plan include a governance model, escalation procedures, and risk criteria. Once a portfolio is established, it is important to monitor and reassess goals annually, making changes as needed. There are some projects and programs which do not directly provide a monetary return but are still necessary. Some examples of these efforts are regulatory compliance, corporate social responsibility, and operational efficiency. Without regulatory compliance checks, the business could be fined and suffer reputational damage or even be shut down. Corporate social responsibility allows the organization brand to be associated with social good and increase revenues through the reputation of this work. (It can also have the inverse affect when done poorly.) Operational efficiency efforts like taking the plant down regularly for maintenance may cost the plant significantly, but planned outages are much more manageable than unplanned ones.

The portfolio will need to select components to achieve those goals, benefits, and deliverables that are defined within the plan. The selection process should take into consideration not only scope, schedule, and cost as expected but also resource requirements, interdependencies, risk and opportunity, complexity, and ability to execute. Just as project documentation undergoes progressive elaboration, so should the portfolio component selection process. Starting with a simpler model that matures over time is a more effective way to manage the process.

One of the most beneficial avenues to begin the maturation process is in the resource management space. Not only does this allow the portfolio to manage its work more effectively, but it can also have personnel benefits. By selecting portfolio components with the portfolio's roadmap and delivery plan balanced against the capability and capacity of the organization, the result of the analysis is the need to bring on additional capacity and/or capability, or to defer work until such time when that capacity and/or capability is available or can be acquired.

Most organizations start the project-selection process by accepting all work that is presented and quickly come to realize that executing all of them is not possible. This ad hoc method is not really a process, but as maturity occurs it moves to a reactionary effort then to a structured process and then to a managed one and eventually to a state of continuous improvement. This concept of a maturity spectrum is consistent with multiple maturity models, including OPM3 from PMI and the portfolio component of CMMI.

Portfolio leaders should start with basic questions like "Are we doing things right?" and "Are we doing the right things?" Additionally, questions to ask include "Where are we going?" and "How do we want to get there?" Then they should implement a ranking system to group the work. The ranking criteria can be random, set, or subject to change based on business or market needs at that moment. Next, implement a scoring system based on business value. The projects with the most strategic value are prioritized for execution. Eventually, selection criteria will be established that include key elements such as resource capability and capacity, funding, risk and opportunity, and value and benefits. It will then be possible to optimize the selection based on strategic value, capacity, and capability of the organization. For example, United Rentals is known in the construction industry for equipment rental, but the organization also has a research and development portfolio where projects are selected in order to deliver one new technical product or feature quarterly. Projects are selected not only for their potential value but also for the time to market. These projects are then packaged for the retail construction market – hopefully before their competitors. This concept takes something very innovative and places strategic parameters on the work to achieve the strategic goals which in this case include maintaining competitive advantage in their industry and expansion into internet-connected construction concepts.

Selecting projects with just a few obvious inputs or simply choosing projects because they are "...the squeakiest wheel..." is not the best technique to choose a portfolio component. These strategies may work sometimes, especially "...when tackling the low-hanging fruit..." but a more structured approach is required when the "...priorities are not so obvious". One of the biggest decisions that any organization would have to make is related to the project investments it undertakes. Once a proposal has been received, there are numerous factors that need to be considered before an organization decides to pursue a project. The most viable options need to be chosen, keeping in mind the strategic goals and requirements of the organization. Some additional variables need to be considered for each of the four disciplines: strategy, finance, risk, and technology (James, 6).

- Benefits and value: A measure of the positive outcomes of the project. These are often described as "the reasons why you are undertaking the project". This should include the enterprise and business value vs. spend ratio.

- Feasibility: A measure of the likelihood of the project being a success, i.e., achieving its objectives. Projects vary in complexity and risk. By considering feasibility when selecting projects, it means the easiest projects with the greatest benefits are given priority (Jordan, 53).
- Solution clarity: Is the solution already known? Does it support the current strategic goals? Does is align with the organizational culture and technology roadmap? (FEA-PMO, 8).
- Availability: Are there organizational resources (funding, staff, and equipment) available to complete the project objectives? How will this project compete with other higher priority projects?
- Stakeholders: Will the successful outcome of the project have a material impact on customers' (internal or external) perceptions of quality or performance?

Most project portfolio management information systems (PPMIS) tools have complex models that are based on the business rules that have been established and a weighted method called Multi-Criteria Decision Analysis (MCDA) which is a valuable tool that applies to many complex decisions. Criteria should be established and weighted including variations that include the following:

- maximizing investment value,
- efficiently utilizing available capability and capacity,
- delivering value against multiple strategic goals and objective, and
- ensuring the portfolio is not comprised of only high-risk projects.

Successful PMOs start governance with project selection by ensuring that strategy alignment is in place before authorizing the project. Well-designed project/program acceptance criteria will impact, at a minimum, four disciplines: strategy, financial, risk, and technical (PWC, 7). Additional areas include time to implement, as well as people and other resources. To ensure effective acceptance criteria for both projects and programs, portfolio leaders should consider the following (see Table 5.4 for illustration):

- Alignment with corporate strategy or being an enabler to deliver corporate strategy
- Financial parameters and milestones (including appropriate KPIs to measure these)

TABLE 5.4

Sample Project Prioritization Matrix for Technology Portfolio

Project Name	Business Value	Ability to Execute	Cumulative Score (40% A2E, 60% BV)	Investment Quadrant	Primary Financial Measure	BU Value	Enterprise Value	Readiness	Technical Characteristics	NPV	Alignment to BU Strategy	Improvement to Ops/Svc	Operational Performance	Retain & Attract Talent	Sustainability	Regulatory & Legislative	Financial Strength	Growth	Sponsorship	Business Resources	Tech Resources	Change Enablement Impact	Project Risk	Endorsement
Business Obj. Replacement	5.2	3.8	4.6	Consider	1.0	9.1	7.0	4.5	3.0	1	10	7	7	1	1	10	3	1	3	7	7	1	3	FALSE
Oracle Upgrade	5.2	9.6	7.0	Go Do	1.0	9.1	7.0	9.3	10.0	1	10	7	7	1	1	10	3	1	7	10	10	10	10	TRUE
Tax System Upgrade	5.8	9.1	7.1	Go Do	3.0	7.9	7.5	8.3	10.0	3	10	3	10	1	1	10	1	1	3	10	10	10	10	TRUE
Ent. Digital Storage	3.9	8.1	5.6	Question	1.0	9.1	2.4	9.3	7.0	1	10	7	3	3	3	1	0	1	10	7	10	10	7	TRUE
Intranet Enhancements	4.1	6.5	5.1	Question	3.0	5.8	3.9	6.0	7.0	3	7	3	3	1	1	7	1	1	3	7	7	7	7	FALSE
SAN Replacement	4.6	7.6	5.8	Question	3.0	7.3	4.0	8.3	7.0	3	10	1	1	1	1	10	0	0	3	10	10	10	7	TRUE
Regulatory Process Enhancements	5.5	7.0	6.1	Go Do	1.0	10.0	7.0	7.0	7.0	1	10	10	7	1	1	7	7	1	7	7	7	7	7	FALSE
HR App. Reader SW	7.1	8.9	7.8	Go Do	7.0	9.1	5.3	7.8	10.0	7	10	7	7	1	1	7	1	1	7	10	7	7	10	FALSE
Internal Email Tracking & Metrics	4.0	7.8	5.5	Question	3.0	7.0	2.4	8.5	7.0	3	7	7	3	3	3	1	1	1	10	7	7	10	7	TRUE
Internal Social Media Updates	3.1	7.4	4.8	Question	1.0	5.8	3.2	7.8	7.0	1	7	3	3	7	1	0	1	0	10	7	7	7	7	FALSE

- Risk parameters including the monetization of risk contingency and adequate management reserve (Jordan, 18)
- Structural architectural changes including the technology roadmap (Jordan, 302)
- Appropriate weighting for each of these considerations by stakeholder tolerances

The same model does not work for every organization. Organizations often make investments based on issues or challenges within the organization. For example, banks and utilities are highly regulated and risk averse whereas technology companies have a much more tolerant approach to risk within their portfolio. Why is risk important? When organizations establish a portfolio-level perspective for managing risk, they are more likely to successfully balance their project portfolio. Strategic project selection should include a portfolio risk assessment because too much risk indicates a dubious organizational strategy with uncertain outcomes and negligible risk could indicate a lack of strategic thinking and missed opportunities. Portfolio risk and opportunity drivers include the following:

- Portfolio Management Plan
- Strategic Goals
- Corporate Culture
- Organizational Strategies
- Executive Commitment
- Change Management
- Stakeholder Management
- People, Process, Systems & Infrastructure Assets
- Key Performance Indicators (KPIs)
- Risk Management Plan

These KPIs should also be tracked over time for trends. A trend analysis chart is a report type that you can use to display the performance of one or more metrics over time. For example, you might use a trend analysis chart to display the historical performance of a KPI and to estimate future results of that KPI. Because KPIs exist as elements in dashboards, a trend analysis chart must use dashboards as its data source (Wise, 1). You can select more than one KPI for your report, but a trend analysis chart can display only one KPI. The following are two suggestions to consider for completing KPI trend charts. First is to annotate the graph with any organizational events like "Reorganization" or "Merger Announcement" so it is easier to explain any anomalies that are detected in the data. Second, do not aggregate the data if at all possible. In aggregate, trends can hide insight and therefore "dirty" the data. If you want to compare "clean" trends, then your best option is to compare different segments within your data.

A methodology is a set of rules associated within a given area of study or discipline. While a methodology does not provide solutions, it offers "best practices" for the work conducted – in this case a project or program. The methodology should not only provide a foundation with rules and policies but also provide flexibility to effectively manage that effort. Examples can be seen in multiple places and industries, and many times symbolize quality expectations such as the International Organization for Standardization (ISO) 9000 in quality management or board certifications within the medical profession. Another truly relevant example is the rules, standards, and principles set forth by the PMI for project, program, and portfolio management practitioners where certifications set a standard and expectations. Having a systematic methodology to follow allows practitioners the opportunity to become more proficient, dependable, and effective as time passes. "According to Aberdeen Research, effective portfolio management can enable companies to achieve up to 25% more revenue from new products when compared with less successful competitors" (PWC, 3).

The rules around governance should lay out the framework for ownership of projects and portfolios, defining the steering committee structure and determining roles and assigning responsibilities. They should define who is authorized to initiate, to continue, to amend, and to stop projects as well as stipulate who sets standards and who controls the overall investment budget (PWC, 3). These rules need to be embodied in a set of governing forums, decision processes, and tools. Process maturity has been found to yield "...more predictable project performance and lower direct project management costs" (MacDougall, 2).

Rules should also provide a basis for acceptable project performance with the rules that govern project milestones, deliverables, and reporting. These rules should define the way to monitor the progress of projects and introduce flexible options to the typical change lifecycle in order to accommodate various kinds of projects. Those rules should also define "... consequences for breaking rules or poor performance" (James, 10). The portfolio manager must be able to influence the behavior of project managers. Rewards and penalties need to tie in with the governance framework and rules. For some companies, this risk and reward concept is implemented for the practitioners by inclusion in their annual performance appraisals. For example, one financial institution tied benefits attainment to PMO performance appraisals by tying the percentage of benefits attainment received to the percentage of targeted annual income adjustment. Another example is a utility that tied project team performance appraisals to safety incidents for its major projects.

Risk methodologies should also be well defined. The PMO should deploy decision tools to help evaluate the risk-adjusted financial performance of projects. PMOs utilizing best practices are starting to use simulation tools to quantify project uncertainty by generating a range of possible economic scenarios to estimate the distribution of project values (Jordan, 30). An example of one such tool is Monte Carlo analysis which provides the probability of outcomes for the scenarios that are input into the tool. It is used in many industries and in government to provide the most and least likely path as well as all of the variations in between. While this particular type of model is used in many diverse types of portfolio analysis, it is particularly useful in the risk management area. These types of models evaluate the often overlooked embedded value in projects by showing the probability of both risk and opportunity. One familiar example of this type of tool can be seen each summer on the weather forecast. Whenever a hurricane has formed, a simulation tool plots the paths the storm will take based on the factors input into the tool. In this case that would include temperature and tides. These simulation tools should have flexibility for usage with smaller projects but be robust enough to manage considerable risk and complexity. This is where benefits-realization success metrics prove their value. Enterprise Program Management Offices (EPMO) and PMOs that continuously track and manage the progress of benefits realization can more easily identify poorly performing projects early on to facilitate the process of managing risk and by doing so the EPMO/PMO can stop projects and programs before wasting additional resources (PWC, 9).

> Beyond governance, an unwavering financial discipline, and regular reviews of portfolio performance throughout the entire process are necessary to guide informed decisions. This level of discipline will demand standardized key performance indicators (KPIs) and powerful analytics to deliver objective insight for proactive decisions.
>
> **(PWC, 3)**

Also essential is a benefits-realization process that enables organizations to ensure projects yield the expected benefits and value so underperforming projects can be stopped early. The process should allow for both financial benefits and strategic value to be measured and monitored.

Portfolio performance itself is the key indicator for effective portfolio governance. With the utilization of portfolio performance tools, the portfolio manager can determine if governance and oversight is effective and identify opportunities to provide further guidance and support (James, 9). Some of these tools should include project/program reviews and portfolio performance dashboards. Project reviews are utilized to identify and help rescue important projects that may be in trouble. These types of reviews may also be scheduled for larger more complex efforts to ensure that they start out and remain on the projected course. A careful assessment can help correct or stop underperforming – and ineffective – projects early.

> The portfolio dashboard should be used to diagnose systematic problems that need to be addressed: e.g., problems with procurement practices, vendor performance, financial management and quality of testing. This approach to portfolio monitoring maximizes the chances of success for projects by providing the best possible development environment (e.g., for procurement, system build and testing).
>
> **(James, 9)**

While the components of the portfolio dashboard vary from company to company, the standard KPIs typically include costs, benefits and value measured against milestones and expectations.

Another aspect of portfolio governance that comes with maturity is benefits realization management. According to PMI, benefits realization management is defined as a "Collective set of processes and practices for identifying benefits and aligning them with formal strategy, ensuring benefits are realized as project implementation progresses and finishes, and that the benefits are sustainable – and sustained – after project implementation is complete" (Benefits, 2). As components are introduced to the portfolio, business cases are named that document potential benefits of the work along with a timeline. This also includes alternative ways to obtain those or similar benefits which is called alternatives analysis. Most organizations do not have any mechanism to ensure that the benefits promised within that business case are ever fully realized. Even if the project or program delivers the work as intended, there are sustainment activities to ensure that the benefits are realized. KPIs and other measures must be in place to effectively manage value delivery and benefits realization. Governance measures should be in place to ensure that these KPIs are met and allow portfolios to be effective by generating value and operational efficiency.

It is critical that organizations employ standardized financial targets as part of business-case reviews. Research shows that financial discipline at the beginning of the portfolio-acceptance process pays off in the long term.

> Organizations with best-in-class project management review the potential for revenue or return 35% more than others before accepting or approving new projects. And when compared with other firms, companies that get portfolio management right tend to assess past performance as an indicator of future efficiency.
>
> **(PWC, 8)**

The process selected should provide the organization with value. While the project itself may not have a high rate of return, it may be the enabler for other projects that will. There are significant factors that should be considered and weighed in a manner that is most beneficial to the organization to achieve its strategy. With each project selection, the question needs to be asked, would this decision help me to increase organizational value?

As projects and programs end their lifecycle, every effort should be made to track the benefits documented in the business case for which the projects were selected; record lessons learned to ensure that future governance can provide assistance to prevent the repetition of poor practices and enable best practices; and transition to operations to realize the benefits and value that were intended for the business.

BIBLIOGRAPHY

Center for Business Practices and PM Solutions Research. (2005). Project portfolio management maturity: A benchmark of current best practices. Retrieved from: http://www.pmsolutions.com/audio/Research-PPM-Maturity.pdf

"Duke Energy IGCC Project." *IN.gov*. Accessed 18 Mar 2022. Retrieved from: https://www. in.gov/oucc/electric/key-cases-by-utility/duke-energy-igcc-project/

Federal Enterprise Architecture Program Management Office (FEA PMO), OMB. (2007). Value to the mission. Retrieved from: http://www.whitehouse.gov/sites/default/files/ omb/assets/fea_docs/FEA_Practice_Guidance_Nov_2007.pdf

Fruhlinger, Josh, Sayer, Peter, & Wailgum, Thomas. "16 famous ERP disasters, dustups and disappointments." *CIO.com* 20 Mar 2020. Retrieved from: https://www.cio.com/ article/278677/enterprise-resource-planning-10-famous-erp-disasters-dustups-and-disappointments.html

James, M., & Ryan, P. (2012). "Effective project portfolio governance". Published by Oliver Wyman. Retrieved from: http://www.oliverwyman.com/insights/publications/2012/ oct/effective-project-portfolio-governance.html

Jordan, A. (2013). *Risk management for project driven organizations: A strategic guide to portfolio, program and PMO success*. Plantation, FL: J. Ross Publishing, Inc.

MacDougal, C., & Cadilhac, C. (2011). Project program and portfolio governance. Building 4 Business. Retrieved from: http://www.building4business.com.au/publish.html

PMI Thought Leadership Series: Guiding the PMO. (2016). "Benefits realization management framework." Retrieved from: https://www.pmi.org/-/media/pmi/documents/public/ pdf/learning/thought-leadership/benefits-realization-management-framework.pdf

PricewaterhouseCoopers, LLC. (PWC). (2012). Strategic portfolio management. Retrieved from: http://www.pwc.com/us/en/increasing-it-effectiveness/publications/strategic-portfolio-management.jhtml

Success in Disruptive Times PMI Pulse of the Profession. (2018). Retrieved from: https:// www.pmi.org/-/media/pmi/documents/public/pdf/learning/thought-leadership/ pulse/pulse-of-the-profession-2018.pdf

The State of Project Management Annual Report. (2019). Wellingtone. Retrieved from: https://wellingtone.co.uk/wp-content/uploads/2019/11/The-State-of-Project-Management-Report-2019.pdf

6

Marketing the Project Portfolio

Rodney Turner

An essential part of communication on projects, programs, and portfolios is their marketing (Turner & Lecoeuvre, 2017). To win the support of the various stakeholders to a project, program or portfolio, the sponsor needs to sell them the benefits, and persuade them that the commitment they are going to make, a price they pay, is worth the benefit to them. There are three organizations involved in the management of projects, programs, and portfolios: the project or program itself; the investor undertaking the project, program, or portfolio, and contractors delivering services to the investor. All three organizations have to do marketing, and they have to market to a range of different stakeholders. The different stakeholders are effectively different market segments, and need messages sent to them, the promotion, tailored in different ways. They will all support the project or program in a different way, and so it will have a different cost to them, and they will all perceive different benefits. So the project or program needs to be promoted to them in different ways and in different places.

The marketing of projects, programs, and portfolios is a developing concept (Turner et al., 2010; Turner & Lecoeuvre, 2017; Turner et al., 2019), and we describe some of the current ideas. In the first section we describe the three organizations involved in project, program, and portfolio management, and the different types of marketing they do. Two of the organizations, the investor and the contractor, do their marketing at the portfolio level. The third, the project or program, markets itself. We then describe the two traditional elements of marketing, market segmentation and the 4Ps. A model of project results suggested by Rodney Turner (2014) is used to suggest a range of different stakeholders involved in projects, programs, and portfolios, and to show how it suggests they need different forms of

DOI: 10.1201/9781003315902-6

communications. We then consider how the different stakeholders perceive the product, price, promotion, and place of sale. We next describe the marketing that needs to be done by the three organizations, the project or program, the contractor, and the portfolio or investor.

THE THREE ORGANIZATIONS

Graham Winch (2014) suggests that there are three organizations involved in the management of projects, Figure 6.1:

1. *The initiator*: is a permanent organization that initiates the project or program, provides the resources for its execution, and receives

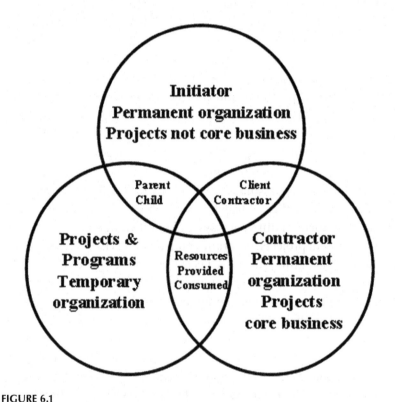

FIGURE 6.1
Three organizations involved in the management and marketing of projects. (© Rodney Turner, see Turner, 2014.)

benefit from the operation of project's output. The investor will manage the projects and programs it does through a set of portfolios, and so undertakes it marketing through the portfolios. This type of marketing we call marketing OF the project.

2. *The contractor*: is a permanent organization that undertakes a portfolio of projects or programs, doing work for a range of clients. Contractors need to do marketing to win new business, and that marketing they also do at the portfolio level. They need to identify where clients have a need for potential new projects, and sell their capability for the project, they need to maintain communication during the project, and then after the project to win new business. This type of we call marketing FOR the project.

3. *The project or program*: is a temporary organization. It is the vehicle through which the investor makes the investment to deliver beneficial change, that is builds a new asset (the project output), which through its operation (the project outcome) will provide benefit. The project needs to engage various stakeholders and so will communicate with the stakeholders to win their support. This type of marketing we call marketing BY the project. (For simplicity, for the rest of the chapter we will refer only to "project" to cover "project or program".)

As you see, two of the organizations involved in project marketing are permanent organizations and do their marketing through the project portfolio. The other is temporary, and the marketing is done at the project level. We will discuss the marketing by the three organizations in the reverse order: project, contractor, and investor; BY, FOR, and OF.

MARKETING

We now consider two elements of traditional marketing as they relate to projects, programs and portfolios:

- market segmentation
- the 4Ps

Case

We are going to illustrate the ideas that follow by suggesting what it means for a project in the United Kingdom: the so-called High Speed 2, HS2. This is a high speed railway line that has been proposed connecting London, Birmingham, and Manchester. It is called HS2 because HS1 was the high speed line that connected London to the Channel Tunnel. HS1 was the first new mainline that had been built in the UK since the late 19th century. All lines built in the 20th century were light rail. Current estimates suggest that HS2 will cost £80 billion, though previous experience on similar projects suggests the final cost may be higher.

Being a mega project, the project is in reality a program, consisting of many lines, bridges and tunnels, stations, and other projects. A program consists of a portfolio of related projects, but also the project is part of a portfolio of rail and other transport projects the government is undertaking.

HS2 has committed a fundamental marketing error in that it has the wrong name. The need for the project is to increase capacity on the West Coast Mainline. There are short distance trains to Watford, medium distance trains to Rugby, long distance trains to Manchester and Birmingham, and freight trains. There is no space for more trains and people between London and Watford have to stand. It is not possible to build new tracks alongside the existing ones. The only way to increase capacity is to build a new line, and if you are going to build a new line you might as well make it high speed. But because the project is called High Speed 2, it creates the narrative that the purpose of the project is to get top Birmingham 10 minutes more quickly. People say quite correctly there is no need to get to Birmingham 10 minutes more quickly, so why are we spending £80 billion? The project needs a name that reflects the purpose, to increase capacity. Also they are taking a modernist approach to marketing, selling the technical; achievement, rather than a post-modernist approach, selling the benefit, that people travelling between London and Watford won't have to stand.

Market Segmentation

Market segmentation is a marketing concept whereby prospective buyers into are placed in groups (or segments) that have common needs and will respond to a marketing action in a similar way. Market segmentation enables companies to target different types of consumers who perceive the

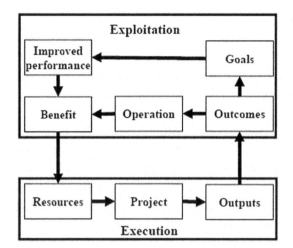

FIGURE 6.2
Results-based view of projects. (© Rodney Turner, see Turner, 2014; Turner, Huemann, Anbari and Bredillet, 2010.)

benefit of certain products and services differently from one another, and so will buy them in different ways and pay different prices for them. It is suggested that each market segment should obey three criteria:

1. Homogeneity: each segment should have common needs
2. Distinction: each segment should be different from the others
3. Reaction: the members of a given segment should react in a similar way and differently from other segments

Figure 6.2 illustrates different levels of project results (Turner, 2014). There are different stakeholders associated with each element, all representing different market segments.

Resources
Resources provided to a project will take many forms: money; materials; people; data. Different stakeholders will be associated with all of them. Bankers may provide money, suppliers materials, and contractors the people. The people actually working on the project are other stakeholders.

On HS2, the project will be paid for by the government out of tax revenues. So the country's tax payers are the ultimate financiers. The government needs to persuade the tax payers that the project is worthwhile. Unfortunately, because the wrong narrative has been created, tax payers are unconvinced.

The Project

There are many stakeholders involved in project execution. But most of them will be covered under the other areas. For instance we have already met the stakeholders providing resources. So we focus here on just the investor, the client, and the main contractor.

The investor: This is the organization proposing the project and will primarily be marketing to other stakeholders trying to win their support for the investment. But the main and other contractors will be trying to win the investor's attention to give them the work of the project. For HS2 the main investor is the government (though tax payers provide the actual money).

The client: The investor is the main client, but in a portfolio or program of projects, there are by definition projects which make up the portfolio of program. Each of those projects may have a client separate from the investor. HS2 will in reality be a program made up of a portfolio of projects. It may be broken into several stretches of line, and bridges, tunnels, and stations, which may be separate projects. There was a high speed line in the Netherlands, where the track foundation was done as a re-measurement contract, with the investor as the client, bridges, tunnels, and stations were done as design and build, again with the investor as the client, but the construction of the track was done as a Public-Private Partnership project, using Build-Own-Operate (BOT), with the client as the Special Project Vehicle that owned the concession agreement. The client needs to market to the investor to win the concession and to the main contractor to encourage them bid for the work.

The main contractor: We met contractors providing materials and labor under resources. The main contractor(s) will manage the design, procurement, construction, and commissioning of the projects that make up the program or portfolio. There may be a different main contractor for each project. They need to undertake the marketing described in Section "Marketing for the Project" of this chapter to win this and future projects.

The Output

The output is the new asset the project constructs. In the case of HS2 it is the track foundation, the track (with signaling), bridges, tunnels, and stations. Many of the stakeholders described under other elements will have an interest in the project's output, so we mention here only the IMBYs (In My Back Yard). There will be NIMBYs (No! Not In My Back Yard), and YIMBYs (Yes Please In My Back Yard). For HS2 the NIMBYs include people whose houses will be demolished or damaged by the construction of the line. YIMBYs may be people who see the project as providing economic opportunity in their community. The investor needs to tell the YIMBYs about the project and the economic opportunity it provides and may do that through television, newspapers, and social media. NIMBYs need to be told how they can claim compensation, and be persuaded that the project is of great benefit to the country, and some collateral damage is unfortunately necessary. They can be contacted in similar ways.

The Outcome

The project's output (the new asset) will be operated to provide new competencies. The people who make use of those new competencies are the consumers. In the case of HS2, the outcome includes:

- Reduced congestion
- Greater capacity meaning people can travel more freely, and will not have to stand between Watford and London
- Faster travel times
- Improved safety with fewer accidents
- Environmental benefits through the switching of traffic from road and air to rail

A post-modernist approach to marketing will tell us about these things. The project is adopting a modernist approach telling us what a wonderful technical achievement it is. Freight companies may be particularly interested in better access to European markets. At project commissioning the investor needs to contact them to make them aware that the new services are there and persuade them to use it. This may involve traditional marketing.

Operation

To provide the outcome, the output is operated. The people doing this are called users or operators. They need to be communicated with throughout the design and construction stages of the project to get their input to the design of the output and its ease of operation. They may also need training. On HS2 they will include train drivers, signalers, station managers and employees and their trades unions.

The Benefit

What the consumers pay to use the service, or the project outcome, provides a benefit to the investor. The consumers also get benefit they are willing to pay for from the operation of the project's output. The company that was investor during the project phases usually becomes the owner of the asset during operation. But sometimes the investor sells the new asset to a new owner who receives the income stream. That is particularly the case for the construction of office blocks or rental housing. An investor identifies the opportunity and constructs the building, and then sells it to a financial institution who will receive the rental income. As we said for HS2, the track may be built as a BOT project, and so the owner of the concession agreement will receive income from the operation of the track. Stations may be sold to financial institutions. Traditional marketing may need to be used to identify them and persuade them to buy the stations.

Goals

With time the operation of the project's output and outcome may enable the investor to achieve higher level goals, and performance improvement. In the case of HS2 this may be economic development and higher tax revenues for the government. But as we said, the ultimate investor in the project is tax payers, and they need to be persuaded that the economic development that makes the project worthwhile. It is the increased capacity and faster journey times to Europe for passengers and freight and the economic development that accrues from it, not getting to Birmingham 10 minutes more quickly, that makes the project worthwhile and tax payers need to be persuaded of this goal.

The 4Ps

Traditional marketing theory talks about 4Ps of marketing: product, price, place of sale, and promotion. Some people now identify seven or even eight Ps, but we want to focus here just on the traditional four.

Product

Marketing theorists will tell you that people do not by physical products, they buy benefits. You do not buy a computer mouse, you buy the ability to control a computer; you do not buy a bottle of water, you buy the ability to quench your thirst. Project managers need to learn that project communication is not about telling people about Gantt charts and critical path networks, but it is about telling investors, consumers, users, YIMBYs, contractors, and others about the benefit they will get from the project. As we saw in the previous sections, the types of benefits they will get will be quite different, and so the messages to the different stakeholders need to be framed in different ways.

Price

Each of these stakeholders will make some commitment to the project, and that commitment is a price they pay to receive their benefit from the project. Some provide financial support, some provide their time, some provide emotional support, and some provide political support. The price people are willing to pay of course depends on the level of benefit they perceive they are getting. So when communicating with project stakeholders, the sponsor needs to understand what benefit they are getting and how they perceive it, and so understand what level of support they are willing to give, and phrase the messages to them accordingly.

Place of Sale

Where will the project impact on the stakeholders and their lives? That is where the communication needs to take place. There will be different ways of communicating with different stakeholders. Contractors will be contacted through the trade press; YIMBYs and NIMBYs through television, newspapers, and social media; consumers via advertising literature; users at their pace of work or through their trades unions.

Promotion

So the message must be tailored to the stakeholder, the benefit they perceive, the commitment they will make, and where the project impacts on their lives. That will be done through the communication plan.

Communication Plan

When developing a communication plan to promote the project to the stakeholders, there are a number of key questions that need to be considered.

Who Is the Target Audience?

There are many different stakeholders; the market for the project needs to be segmented, to understand what the objectives of each stakeholder are, what benefit they perceive, and what contribution they are going to make.

What Are the Objectives of Each Communication?

Recognizing the objectives enables the people doing the promoting to understand what they are trying to achieve when they communicate with each stakeholder.

What Are the Key Messages?

Once the objectives are known, the information that needs to be communicated to each stakeholder can be identified.

Who Will Do the Communicating?

Different stakeholders need to be communicated with by different project participants. Who does the communicating may also depend on the information being communicated. The communication may be done by the investor, the main contractor, the portfolio manager, the program manager, the project manager, or technical experts. To be credible, some information needs to come from senior managers, other information from technical experts, and yet other information from people who understand project progress.

When Will the Information Be Given?

Different stakeholders need to be contacted at different times. The YIMBYs and NIMBYs may need to be first contacted before the feasibility study starts, to win their support from an early stage. Financiers and the main contractor need to be contacted during the feasibility study. The first contact with the consumers needs to be made then to demonstrate the commercial viability of the project and to the users to show technical feasibility. Suppliers of labor and material are first contacted during design, and the consumers and users are recontacted during commissioning. Communication with financiers needs to be maintained during operation to show their investment is being repaid. The National Aeronautics and Space Administration (NASA), for example, continues to market Project Curiosity, the Mars rover, during its operation to maintain support of the public and politicians.

How Will Feedback Be Encouraged?
Communication is two way. Communication is not transmission on information. You do not send out a message and hope somebody hears it. Do not scream in space. Under communication a message is sent, someone hears it and sends a response, and the response is heard confirming the message has been understood. So feedback must be encouraged. Ask people to respond, and show you are listening, through your body language, by answering their questions and by acting on their suggestions. If a stakeholder makes a suggestion that increases the value of the project, do it. If they make a suggestion which reduces the value of the project, avoid it. But if they make a suggestion that has no impact the value of the project, it may still be worthwhile implementing it to show you are listening.

What Media Will Be Used?
The following media can be used:

- Social media: Facebook, Twitter, TikTok, Instagram, You-tube and Linked-In
- Project web pages
- Newspapers and radio
- Seminars, workshops, focus groups, control meetings
- Videos, CDs, and newsletters
- Bulletins, briefings, press releases
- Project, program and portfolio lunches
- Open days and project exhibitions

MARKETING BY THE PROJECT

The first organizations we consider is the project itself. This is marketing by the project to engage stakeholders. Figure 6.3 is a stakeholder engagement process (Turner, 2014).

Emotional Intelligence

Before I describe this process, and how it relates to project marketing, I introduce a model of emotional intelligence. Marketing and the engagement of stakeholders requires emotional intelligence. Figure 6.3 is

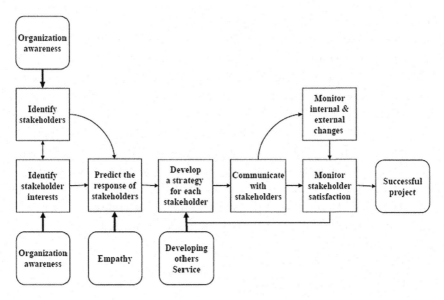

FIGURE 6.3

Stakeholder engagement process. (© Rodney Turner, see Turner, 2014.)

a model of emotional intelligence suggested by Ralf Müller and Rodney Turner (2010). There are four major competences comprising 19 sub-competencies, Table 6.1. The objective of emotional intelligence is relationship management, which is the objective of stakeholder engagement, building relationship with the stakeholders to win their support for the project, by persuading them that the benefits they will get are worth the commitment they will make. The four competences of emotional intelligence are:

1. *Self-awareness*: The leader needs to be aware of his or her own emotional responses to situations. When he or she is self-aware, two more competences follow.
2. *Self-management*: When the leader is aware of his or her emotional responses, he or she can manage the responses.
3. *Social awareness*: He or she can then be aware of emotional responses in other people. From this and self-management flows the fourth competence.
4. *Relationship management*: When the leader can manage his or her own emotional responses and is aware of responses in other people, he or she is better at building relationships with the other people.

TABLE 6.1
The 19 Competencies of Emotional Intelligence

Competence	Associated Competencies
Self-awareness	Emotional self-awareness
	Accurate self-assessment
	Self-confidence
Self-management	Emotional self-control
	Adaptability
	Initiative
	Optimism
	Transparency
	Achievement
Social awareness	Empathy
	Organizational awareness
	Service
Relationship management	Building bonds
	Teamwork and collaboration
	Inspirational leadership
	Influence
	Developing others
	Conflict management
	Change agent

The 19 competencies associated with these four competences are shown in Table 6.1. We mainly focus on those associated with social awareness and relationship management.

The Stakeholder Engagement Process

Identify the Stakeholders and Their Interests
The first step is to identify the stakeholders and their interests. Figure 6.2 and the ensuing discussion should help in this step. This requires organizational awareness: you need to be aware of who the potential stakeholders are, and of their interests. You are trying to identify what commitment they have to make to the project (price) and what benefit they will perceive from the project (product). You also need to be aware of how the project will impact on their lives (place of sale).

Predict the Response of Each Stakeholder
You need to try to understand how each stakeholder will respond to the project. What benefit, if any, do they perceive from the project, what commitment do they have to make, and if so how will they respond. This

continues to require organizational awareness but also requires empathy; you need to be able to predict their responses. There are three questions you can ask about each stakeholder to help in this step:

1. Is the stakeholder for or against the project, or ambivalent?
2. Can the stakeholder influence the outcome?
3. Is the stakeholder aware of the project and its outcomes?

Different stakeholders will have to make different levels of commitment, perceive different benefits, have different levels of influence, and different levels of awareness. So the answers to these questions will identify different market segments for stakeholders in the project.

Develop a Communication Plan for Each Stakeholder
We come to the fourth P, how will the project be promoted? This continues to require organizational awareness to understand what motivates the stakeholders, and empathy, to be able to predict how they will respond. However, you also need skills of developing others and service. You need to develop people so that they buy into the project and can make a positive contribution to it, but at the same time you are providing them with a service, the benefit the project gives them, and you need to be aware of that and develop you own sense of service. We described in the previous sections the components of a communication plan.

Communicate with Each Stakeholder
Then you need to enact the communication plan with each stakeholder. Through this active approach, you will be building bonds and providing inspirational leadership.

Monitor Stakeholder Satisfaction
As the project progresses, you will monitor stakeholder satisfaction. This requires organizational awareness and empathy. You must have empathy to sense how stakeholders are reacting. You will also need to continue to build others and provide inspirational leadership. However, you will also need to influence teamwork and collaboration, to encourage stakeholders feel they are working on a single team toward a shared goal. Through that you will also need to influence their behaviors. If stakeholders are not behaving as expected, you may need to revisit your communication strategy.

Monitor Internal and External Changes
It is also possible unexpected changes occur, which change the stakeholders' positions and thus their response to the project. These need to be monitored, and if they mean that stakeholders are not behaving as expected, then you may need to revisit your communication strategy. It is possible that the changes will create conflicts and so draw on your conflict management skills.

Successful Project
Hopefully, the stakeholders engage with the project as desired, leading to a successful project and proving your skills as a change agent.

MARKETING FOR THE PROJECT

The second organization we consider is the contractor. This project-based organization needs to win the next and future projects, and marketing at this level needs to be conducted as part of the portfolio of projects it is doing for all its clients, because as we shall see some marketing activities take place before, after, and between projects [Turner et al. (2019)].

Marketing Models

An early model was developed by Cova, Ghauri and Salle (2002), Figure 6.4. They suggested three phases of project marketing:

Independent of any project: Before any project has been actually identified, the contractor should be in contact with potential clients, and make themselves aware of the nature of their business and the types of projects they can do. They should make their potential clients aware of their competencies, and make the client aware of how they (the contractor) can satisfy the client's potential needs. The contractor also needs to be aware of any projects the client may be considering, and try to influence the client's definition of those projects so the projects match their competencies and not those of their competitors.

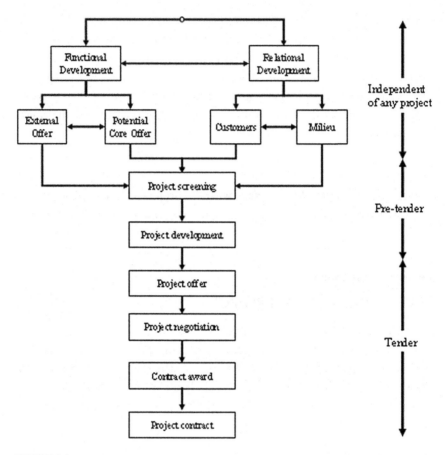

FIGURE 6.4
A three-stage project marketing process as viewed by the project marketing literature. (After Cova et al., 2002.)

Pre-tender: Once the contractor has identified a potential client, the contractor must liaise with them to determine whether it is worth bidding for the project. The contractor must truly understand the client's requirements and whether they match its (the contractor's) competencies. The contractor may continue to try to influence the definition of the project to make it more closely match its competencies. The contractor also needs to understand the chance of winning the bid, and determine whether the potential profit is justified by the cost of bidding, and the chance of winning. As we describe in the subsequent sections, throughout this stage and the next, the contractor needs to market to three types of decision makers in the client organization: the strategic decision makers, the operations managers, and the technical managers. They may require three teams of people to market to each of these three groups (Turner et al., 2019).

Tender: If the contractor decides to bid, they must prepare the tender, make the offer, engage in the negotiations, be successful as the contract is awarded, and start the work of the project.

Throughout these three stages, the emotional competencies identified in this chapter are important, but particularly organizational awareness, empathy, service, building bonds, inspirational leadership, and influence.

Laurence Lecoeuvre-Soudain and Philippe Deshayes (2006) developed a four-stage model. They combined independent of any project and the initial stages of pre-tender into pre-project. They combined the latter stages of pre-project, tender preparation, and mobilization into project start. They then introduced two new stages: project delivery and post-project. During the project, the client's decision makers are project stakeholders, and they need marketing as described as described in the previous sections. But the key new insight is post-project. The contractor must remain in contact with the client looking forward to future projects. This raises project marketing to the portfolio level, since contact with clients is maintained through the portfolio of projects the contractor is doing. If the contractor is providing logistical support to past projects, it helps the contractor maintain contact with the operational managers, but they must also maintain contact with strategic decision makers and technical managers.

The Focus of Project Marketing

Laurence Lecoeuvre-Soudain and Philippe Deshayes (2006) suggest six areas of focus of project marketing, Figure 6.5:

1. Relationship management (Rel)
2. Trust (Tru)
3. Collaboration (Col)
4. Communication (Com)
5. Training (Tra)
6. Going with (providing mentoring, coaching and support) (Gwi)

Relationship management, trust, and collaboration we mentioned previously as emotional competencies, and communication we mentioned as a

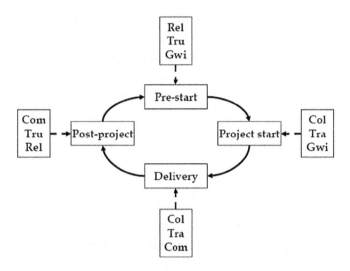

FIGURE 6.5

A four-stage project marketing process as viewed by the project marketing literature. (© Rodney Turner, adapted from Lecoeuvre-Soudain and Deshayes, 2006.)

key element of marketing. Training and going with are closely related to the emotional competency of developing others. Figure 6.5 shows how the emphasis on these six areas of focus varies during the four stages of project marketing. Turner et al. (2019) explored these further.

1. *Pre-project*: The emphasis is on building relationships and trust, and on providing mentoring coaching and support.
2. *Project start*: The emphasis is on collaborating and training and continuing to provide mentoring, coaching, and support.
3. *Project delivery*: The emphasis is on collaboration, training, and communication.
4. *Post-project*: The emphasis is on communication to maintain relationships and trust.

Who Is the Focus of the Contractor's Marketing

Figure 6.6 suggests that the contractor needs to aim its marketing at three sets of decision makers in the client organization (Turner, 1995). The contractor should identify different groups of people to target these three groups of people (Turner et al., 2019).

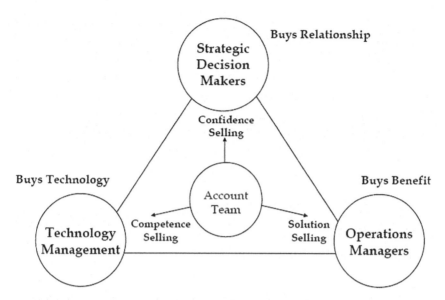

FIGURE 6.6
Three customers for the contractor's account team. (© Rodney Turner, see Turner, 1995.)

The strategic decision makers: These are the people who will ultimately decide to do the project, and determine which contractor will be awarded the contract. These people are interested in the project's goal, Figure 6.2. The contractor's board of directors should target these people, with the help of the marketing department.

The operations managers: They are both the users and the consumers. There may be one set of operations managers who will operate the project's output (the users), and another set who will make use of the project's outcome, and obtain the benefit (the consumers). These people are not interested in the technology. The consumers want the project's outcome to satisfy their requirements and provide them with adequate benefit. The users want ease of operation. It will usually be the role of the sales and marketing department to communicate with these people, though the project manager may also be involved. It is essential to make them comfortable that the project's outcome will satisfy their requirements and provide them with the benefit they want, and that the output will be easy to operate.

The technical mangers: These are the people who will judge the contractor's technical solution and will be able to determine whether the project's output will work to provide the outcome. The contractor's technical managers must communicate with these people to persuade them of the contractor's technical competence.

Many years ago, I went on a skills of persuasion course. We were shown videos of John Cleese illustrating mistakes made by vendors. In one he is a salesman selling a photocopier. He is singing the praises of the fancy electronics in the photocopier to the potential buyer. The buyer looks somewhat non-plussed and says, "I don't want fancy electronics; I want clear copies." Contractors however have to sell the fancy electronics to the technical managers, and the clear copies to the operations managers.

MARKETING OF THE PROJECT

The third organization involved in the management of projects is the investor, the organization that identifies the project opportunity, that an asset can be built, or a change undertaken that will provide benefit, initiates to the work of the project to build that asset and sources the finance. Very little has been written about this element of project marketing. Figure 6.7 shows the project as part of an investment process (Turner, Huemann, Anbari and Bredillet, 2010). The project builds the output, the investment, a new asset that will be operated to provide benefit to repay the cost of its construction. The project is part of the investment process. Figure 6.8 is a typical project process with six stages: concept, feasibility, front-end design, execution, close-out and exploitation. In Figure 6.7, pre-project covers concept and feasibility in Figure 6.8, independent of any project and pre-tender in Figure 6.4, and pre-project in Figure 6.5. In Figure 6.7, project covers frontend design, execution and close-out in Figure 6.8, it includes tender in Figure 6.4, and it includes project start and delivery in Figure 6.5. In Figure 6.7, operation and maintenance covers exploitation in Figure 6.8, and includes post-project in Figure 6.5. In Figure 6.7,

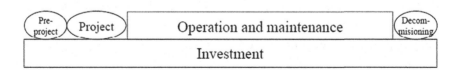

FIGURE 6.7
The investment process. (after Gareis, 2005; see Turner, Huemann, Anbari and Bredillet, 2010.)

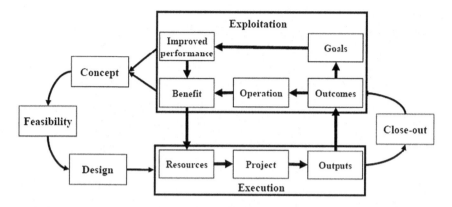

FIGURE 6.8
The project and investment process and overlaid on the three levels of project results. (© Rodney Turner, see Turner, 2014).

decommissioning is not covered by any of the other models. The investor needs to market the project, and its output, outcome, and goals at all the stages of the investment process. That is concept, feasibility, front-end design, execution, close-out and operation as shown in Figure 6.8, and also decommissioning as shown in Figure 6.7.

Early work on marketing during the pre-project and early project phases was done by Susan Foreman (1996). However since she was looking at internal marketing, she mainly focused on one part of the initiating organization marketing the project to other parts of the same organization, and so to an extent it fell under stakeholder engagement. Marketing of the output itself, or the product made by it during commissioning and early operation, can be covered by standard marketing theory. However we believe different marketing approaches need to be used during the pre-project and early project phases, and marketing during the operation phase to win political or public support for future projects, as shown in Table 6.2.

> *Concept*: The sponsor identifies a problem or opportunity and suggests a new asset can be built to solve the problem or exploit the opportunity so that the expected benefit will justify the expected cost. A potential investor needs to be persuaded to support the concept by showing them the investment will provide adequate returns. They need to be persuaded to provide finances for the feasibility stage against the potential benefits. Political or public support needs to be obtained for

TABLE 6.2

Marketing of the Project's Output by the Investor

Project Stage	Aim of Marketing	Examples
Concept	Win support for the initial project concept Obtain political and public support Gaining the interest of the potential investor Obtain funding for feasibility	Winning public support for infrastructure projects. For instance with HS2, persuading that: 1. Travelling to Birmingham 10 minutes more quickly is worth £50bn 2. Economic regeneration of the north of England is worth £50bn
Feasibility	Maintain support for the project concept Maintain political and public support Obtain funding for front-end design	
Front end-design	Convince interested parties in the efficacy of the proposal Maintain political and public support Obtain funding for detailed design and execution	
Execution	Maintain political and public support	
Close-out	Sell the project's output Establish a market for the outcome	Sell a newly built building to an investment company or to the owner occupiers Develop a market for a new product
Operation	Establish a market for the product made by the project's output Prove to the public or political sponsors the value of the product and obtain support for future projects	Use standard marketing For example NASA's marketing of Project Curiosity, the Mars rover

infrastructure projects. Recall the example in the United Kingdom is the HS2 project, to build a high speed line to the North of the country. The wrong message has been sold. People do not think it is worth £80 billion to reduce the journey time to Birmingham by 10 minutes. They might be better persuaded it is justified by the potential economic development, and to save standing all the way to Manchester.

Feasibility: The feasibility of the concept is shown, and the initial business plan is developed. The interest of the investor needs to be maintained, and the investor needs to be persuaded to provide finance for front-end design by being shown the potential returns justify the cost. Political and public support needs to be maintained for infrastructure projects.

Front-end design: The business plan is further developed. The investor needs to be persuaded to provide the finance for execution, by being further persuaded of the efficacy of the proposal to provide adequate returns. Political and public support needs to be maintained for infrastructure projects.

Execution: As money is spent the investor may waiver in its commitment. It needs to be maintained. If the project does not progress as expected, especially if it becomes overspent, political and public support must be maintained.

Commissioning: A developer will try to sell the asset. Otherwise an internal or external market needs to be developed for the outcome. In the case of the developer or external marketing, it will be covered by standard marketing theory. In the case of the internal market it will be covered by the work of Susan Foreman (1996).

Operation: If the outcome is being sold, the market for that product must continue to be developed. Also the value of this new asset may need to be demonstrated to win political or public support for future projects.

CLOSING REMARKS

Communication is an essential part of project, program and portfolio management. Ralf Müller and Rodney Turner (2010) showed that project and program managers are not very good at communication, but it is correlated with success on all projects. A primary focus of marketing for the project and its parent organization is to win support for the project, its output and outcome, and we have shown that is covered by the concepts of marketing. We have also shown that support for the project is an emotional commitment by the stakeholders, and so the emotional intelligence of the person doing the communication is critical. The contractor is not trying to win support for the project, its output and outcome, but its involvement in the project and the focus of marketing are different.

REFERENCES AND FURTHER READING

Cova, B, Ghauri, P & Salle, R (2002). *Project marketing – beyond competitive bidding.* Chichester: Wiley.

Foreman, S (1996). Internal marketing. In Turner, JR, Grude, KV, & Thurloway, L (eds) *The project manager as change agent.* London: McGraw-Hill.

Gareis, R. (2005). *Happy Projects.* Manz.

Lecoeuvre-Soudain, L & Deshayes, P (2006). From marketing to project management. *Project Management Journal 37*(5):103–112.

Müller, R, & Turner, JR (2010). *Project oriented leadership.* Aldershott: Gower.

Turner, JR (1995). *The commercial project manager.* London: McGraw-Hill.

Turner, JR (2014) *The handbook of project-based management* (4th ed) New York: McGraw-Hill.

Turner, JR, Huemann, M, Anbari, FT, & Bredillet, CN (2010). *Perspectives on projects.* New York: Routledge.

Turner, JR & Lecoeuvre, L (2017). The marketing of organizational project management. In Sankaran, S, Müller, R, & Drouin, N (eds). *Organizational project management: Achieving strategies through projects.* Cambridge: Cambridge University Press.

Turner, JR, Lecoeuvre, L, Sankaran, S, & Er, M (2019). Marketing for the project: Project marketing by the contractor. *International Journal of Managing Projects in Business, 12*(1): 211–227.

Winch, GM (2014). Three domains of project organising. *International Journal of Project Management. 32*: 721–731.

7

Portfolio Management Success

Carl Marnewick

INTRODUCTION

Kendall & Rollins (2003) state that there are four major reasons why portfolios are unsuccessful. These reasons are as follows: (i) too many projects in the portfolio, (ii) the wrong projects in the portfolio, (iii) projects not linked to the strategy of the organisation, and (iv) the unbalanced portfolio.

Portfolio success is measured in terms of the aggregate investment performance and benefit realisation of the portfolio (Project Management Institute, 2013b). This definition from the Project Management Institute (PMI) implies that the success of a portfolio is measured over an extended period and that the success is also linked to the strategic intent of the organisation itself. But to understand and analyse how the success of a portfolio is managed, it is important to understand what portfolio management entails. Portfolio management is focused on the achievement of the organisational strategies and objectives. The implementation of organisational strategies and objectives might take anything from months to years. In the case of large corporate organisations, the emphasis is more on years rather than on months. It also implies that to measure the success of a portfolio remains difficult as the strategies and objectives span across the entire organisation and all its divisions. The success of a portfolio is therefore complex and integrated.

The success of a portfolio is also measured against the benefits that need to be realised. Each and every project and/or programme that forms part of a portfolio will have an associated business case that promises some benefits. It is these benefits that need to be realised through the portfolio as these are only realised after the closure of a project and/or programme.

DOI: 10.1201/9781003315902-7

Jonas (2010) states that it remains difficult to capture the overall success or failure of a portfolio. This might be because portfolios are dynamic, multiple interdependent systems that constantly change and develop. There is a need for a comprehensive success framework that is capable to cover the portfolio in its entirety and additionally takes into consideration that changes made within a portfolio will take some time to have either a positive or negative effect. Portfolio success is therefore realised at different points during the lifespan of a portfolio (Jonas, 2010).

Literature also suggests that portfolio success should also be examined multi-dimensionally on the single project, portfolio, and organisational level (Blomquist & Müller, 2006; Müller, Martinsuo, & Blomquist, 2008). In contrast to the PMI definition, portfolio success is defined by (i) the average project success over all projects, (ii) the exploitation of synergies between projects within the portfolio that might additionally increase the overall portfolio value, (iii) the portfolio fit to the organisation's business strategy, and (iv) the portfolio balance in terms of risk, area of application and use of technology (Beringer, Jonas, & Kock, 2013).

Therefore, to assess a portfolio and its positive or negative effects on the organisation, the results have to be measureable and have to cover a wider perspective than the isolated project (Meskendahl, 2010). This raises the question of what constitutes the success criteria of a portfolio. The next section focuses on four major criteria, and it must be noted that they can be extended by organisations.

PORTFOLIO SUCCESS CRITERIA

The discipline of portfolio management owes its origins to a seminal paper written in 1952, in which Harry Markowitz laid down the basis for the Modern Portfolio Theory (MPT). MPT allows determining the specific mix of investments generating the highest return for a given level of risk (De Reyck et al., 2005).

The main portfolio success criteria according to Beringer et al. (2013) and Meskendahl (2010) are as follows:

1. Maximisation of the financial value of the portfolio,
2. Linking the portfolio to the organisation's strategy,
3. Balancing the projects within the portfolio taking into consideration the organisation's capacities,

FIGURE 7.1
Portfolio success criteria.

4. The average single project success of the portfolio, and
5. Sustainability

These criteria are mutually inclusive of each other. Figure 7.1 is a graphical presentation of these criteria and shows the interdependency between the four criteria.

Bannerman (2008) follows the same reasoning but provides no detail on how the portfolio should be assembled. He also states the business objectives that motivated the investment must be achieved. The achievement of these business objectives will lead to the success of the organisational strategy. The achievement of the business objectives is achieved through (i) the various business cases which must be validated throughout the lifespan of the project and (ii) the subsequent business benefits that must be realised.

Financial Value of the Portfolio

The first major success criterion that a portfolio is measured against is that the portfolio must maximise the financial value of an organisation. The song by R. Kelly, *"Money makes the world go round,"* summarises one of the reasons why organisations do exist.

Organisations are there to create profits for their owners and shareholders. The same applies to non-profit organisations. Although these organisations are not focusing on profits per se, they should still be sustainable, and this can only be achieved if the organisation makes a profit or is financially viable.

The aim of portfolio management is to diversify investments in such a way that it reduces the total risk of a portfolio, but this must be done in an effective way and manner. In the realm of portfolio management, the aim is to optimise the total financial value of all the projects within a portfolio, but at the same time, risk exposure should be minimised. There are various ways to measure the financial value of a portfolio, but it is not the purpose of the chapter to elaborate on these measures. The portfolio manager must make a clear distinction between "financial portfolio" management and "project portfolio" management.

The following measures can be used to determine the overall financial value of a portfolio (Reilly & Brown, 2012):

1. *Treynor Portfolio Performance Measurement* measures the returns earned in excess of that which could have been earned on an investment that has no diversifiable risk. This measurement does not take the diversification of a portfolio into consideration.
2. *Sharpe Portfolio Performance Measurement* is the same as the Treynor measure, but the focus is on the total risk of the entire portfolio. It examines the performance of an investment by adjusting for its risk. The ratio measures the excess return per unit of deviation in an investment asset or a trading strategy, typically referred to as risk.
3. *Information Ratio Performance Measurement* measures the average return of a portfolio in excess of a benchmark portfolio divided by the standard deviation of this excess return.

The portfolio manager must examine all the financial parameters and determine which financial factors should be used. Financial factors include investment commitment, return on investment, and the investment period itself (Hill, 2007).

Strategic Success

Strategy maps are used to describe the vision and strategies of the organisation by means of processes and intangible assets. Strategy maps can be used to align intangible assets such as information technology with the organisational strategies and ultimately the vision of the organisation. It is the duty of the portfolio manager to align all the portfolio components to the strategy of the organisation. One way of doing it is through the use of strategy maps and balanced scorecards (Kaplan & Norton, 2004).

A strategy map starts with a vision and follows a top-to-bottom and a bottom-to-top approach. This approach means that the vision dictates all the lower levels, but the bottom-up approach enables the organisation to link everything to the vision. Although the approach is a top-to-bottom approach, the lower levels must be linked to the upper levels to ensure that there is a consistent link between the upper and lower levels of the strategy map. The top-to-bottom flow enables an organisation to take the vision and break it down into its different components and ultimately into different projects. The purpose of the strategy map is to take the vision and break it down into measurable components. It must be made clear that not all the components need to be incorporated into the strategy.

The strategy map provides a framework for an organisation, but it is up to the organisation to determine which of the components it wants to use. The four perspectives of (i) financial, (ii) customer, (iii) internal, and (iv) learning and growth provide the different focal points of the strategy map. The strategy map makes use of the four perspectives. The financial and customer perspectives provide the strategies of the organisation, whereas the internal and learning and growth perspectives provide the business objectives associated with the strategies.

This criterion represents the highest level of benefit achieved by a project, despite the possibility of failures against lower level criteria, as recognised by external stakeholders, such as investors, industry peers, competitors, or the general public, dependent upon the nature of the project (Bannerman, 2008).

For each of the projects identified with a business objective, a business case must be drafted to ensure accountability.

Business Case
The business case describes the justification for the project in terms of the value to be added to the business as a result of the deployed product or service (International Institute of Business Analysis, 2009). The purpose of the business case is to determine whether or not an organisation can justify its project investments to deliver a proposed business solution. Bradley (2010) states that the business case is a living document that needs to be constantly updated throughout the project life cycle. This update will happen at certain point within the lifespan of the portfolio and will determine the continuous inclusion of the project in the portfolio. The business

case must drive the project activities and is used to determine whether the project is still desirable, viable, and achievable.

A business case exists to ensure that, whenever resources are consumed, this supports one or more business objectives. The implication is that a business case must be reviewed at the various stages during the project life cycle. Business cases are developed but are solely used to obtain funding approval for the huge up-front financial investment and not to actively manage the project (Eckartz, Daneva, Wieringa, & Hillegersberg, 2009).

It is a common practice in organisations to approve projects based on a business case. Yet, it is indicated through research done by Eckartz et al. (2009) that many organisations are not satisfied with their business cases. Cooke-Davies (2005:2) shows that many organisations find it difficult to state that projects are "approved on the basis of a well-founded business case linking the benefits of the project to explicit organization goals (whether financial or not)." Many organisations are also unable to state that they had a means of measuring and reporting on the extent to which benefits have been realised at any given point of time.

The Association of Project Management (APM) Body of Knowledge states that the business case *"provides justification for undertaking a project, in terms of evaluating the benefit, cost and risk of alternative options and rationale for the preferred solution."* (Association for Project Management, 2006:129). The Office of Government Commerce states that the business case must drive the project (Office of Government Commerce, 2003). Any project should not be started if there is not a satisfactory business case. In the case of PRojects IN Controlled Environments (PRINCE2), the business case is defined as the reasons for the project and the justification for the project, based on the costs, risks, and expected benefits. PMI states that the business case determines whether the project is worth the investment from a business point of view (Project Management Institute, 2008).

Any decision taken about the inclusion of a project into the portfolio should take into consideration the promised benefits as stated in the business case.

Benefits Realisation
Benefits should be identified and quantified before a project is initiated (Remenyi, Money, & Bannister, 2007). For a project to be judged a success, potential benefits need to be identified as early as possible and realised during the lifespan of the portfolio (Remenyi & Sherwood-Smith, 1998). Benefit identification is the first step of the benefits management process

and identifies and documents benefits that will be most relevant and convincing to decision-makers (Bennington & Baccarini, 2004). In general, it can be stated that the proposed benefits from a project must link in some way to the objectives of the organisation itself (Dhillon, 2005).

Williams & Parr (2008) agree that the process of benefits management begins with the identification of benefits before a programme is initiated and continues with the measurement of the benefits even after the programme has been delivered. It can therefore be concluded that in both programme and project management, the benefits must be defined before the programme and project are initiated. The implication is that it is a function within both disciplines and is not mutually exclusive to either.

Benefits associated with a project are major determinants in its selection and funding. The selection of projects is difficult because there are various quantitative and qualitative factors to be considered such as organisational goals, benefits, project risks and available resources (Chen & Cheng, 2009). The main reason that benefits are identified and quantified seems to be the need to gain project approval (Dhillon, 2005).

This implies that well-structured project selection criteria will help ensure that organisations select projects that will best support organisational needs. It further identifies and analyses risks and proposed benefits before funds and resources are allocated (Stewart, 2008). However, often benefits are primarily strategic or tactical in nature, and their financial rewards are difficult to forecast. The purpose of a formal project selection process is to weigh the risk and rewards of each identified project based on whether it is ideal and pragmatic to initiate (Williams & Parr, 2008).

The selection of projects based on its benefits is directly linked to the strategic objectives that guide the conceptualisation and selection of initiatives and/or projects. The ultimate benefit for any organisation is the realisation of its vision and strategies.

Selection and Balancing

The selection and continuous balancing of the projects that constitute the portfolio is the third success criterion. Projects should and must be selected based on the overall contribution to the achievement of the organisational strategies and objectives.

There are various ways and manners to select the components of a portfolio. Table 7.1 highlights the steps involved in the selection of portfolio components.

TABLE 7.1

Selection of Portfolio Components

	ISO 21502 (Guidance on Project and Programme Portfolio Management)	PMI's Standard on Portfolio Management (2013)
1	Define the contribution of each component to the strategies and objectives	Identify the components
2	Rank the contribution of each component to the strategies and objectives	Categorise the components
3	Define the exposure to risk of each component	Score the components
4	Rank the exposure to risk of each component	Rank the components
5	Define the resource constraints	

The important aspect is that each component must be evaluated on a quantitative as well as qualitative manner. The portfolio manager can use a variety of criteria to define the contribution of each component. Criteria that can be used include the following:

1. Alignment to the strategy
2. Alignment to the objectives
3. Benefits which can be defined as financial and/or non-financial
4. Market share
5. Risk exposure
6. Regulatory compliance

Based on the pre-defined criteria, a weighted scoring model can be used to determine each component's contribution to the portfolio as per Table 7.2.

TABLE 7.2

Weighted Scoring Model for Portfolio Component Selection

	Weighting	Component 1		Component 2		Component 3	
		Score	Total	Score	Total	Score	Total
Strategic Alignment	20	80	4	40	8	70	14
Objective Alignment	20	56	4	90	18	75	15
Benefits	20	70	4	90	18	50	10
Market share	10	80	1	65	6.5	80	8
Risk exposure	15	90	2.25	70	10.5	45	6.75
Regulatory compliance	15	30	2.25	30	4.5	60	9
TOTAL	100		17.5		65.5		62.75

Once each component's contribution has been determined, the next step is to select the appropriate components based on resource constraints. This is also where the balancing of the portfolio plays a major role. Organisations want to do all projects, but only certain projects can be done at a certain point in time based on the available resources. Resources that might have an impact on the amount of projects that form part of the portfolio include the following:

- The amount of human resources available to implement the various projects
- The financial status of the organisation
- Risk exposure

The ultimate goal is to have an optimised portfolio of projects that will be implemented within the constraints of the organisation. A master schedule of resource allocation is necessary to plan the consolidated demand for portfolio resources (Project Management Institute, 2013b).

Project Success

The fourth main criterion that contributes to portfolio success is the success of the individual projects itself that form part of the portfolio. The focus within portfolio management is that each project must be a success in its own right, but at the end must and should contribute to the overall success of the portfolio.

For example, a project may have delivered a product that functionally exceeds the customer's expectations, but in its development it has exceeded the schedule and cost constraints of the accepted scope of the project. Should this project then be classified as a failure or a success? An example that comes to mind is the Sydney Opera House that was completed ten years late and overbudget by more than 14 times. But today it is an iconic site.

It is important that the concept of project success be defined in general as well as specifically for an organisation within a specific industry. Unfortunately, literature remains vague regarding project success. Project success is defined by the (Project Management Institute, 2013a) as the quality of the product and project, timeliness, budget compliance, and the degree of customer satisfaction. According to the Projects and Program Management for Enterprise Innovation (P2M), "a project is successfully completed [when] it delivers novelty, differentiation and innovation on its product, either in a physical or service form" (Ohara, 2005:16). The APM Body of Knowledge defines project success in a similar fashion, stating that it is project stakeholders' needs that must be satisfied, and this is measured by the success criteria as identified and agreed upon at the start of the project (Association for Project Management, 2006). PRINCE2, on the other hand, does not explicitly define project success but states that the objectives of the project need to be achieved (Office of Government Commerce, 2003).

Figure 7.2 is a summary of the success criteria of these major project management standards and methodologies.

Figure 7.2 paints a picture of several criteria contributing to project success. In order to understand project success in totality, Hyväri (2006) suggests that the critical success factors (CSFs) must also be determined.

FIGURE 7.2
Project success criteria. (Marnewick, 2012)

If the CSFs are in place, then project success should follow as a natural outflow.

Hyväri (2006) suggests five CSFs and each of these have sub-CSFs. These CSFs range from the project itself (size, end-user commitment) to the environment (competitors, nature, social environment). If companies and project managers focus on these CSFs, then project success should be assured. Dekkers & Forselius (2007) highlight the importance of scope management as a CSF where project managers can learn and embrace proven approaches that measure the size of software projects, streamline the requirements articulation and management, and impose solid change management controls to keep projects on time and on budget.

Projects and their subsequent products and/or services cannot be seen in isolation. According to Bannerman (2008), the success of the project can be measured on five levels: (i) process, (ii) project management, (iii) product, (iv) business, and (v) strategy. A project might deliver a service late and over budget, but it still delivers on the company's strategy. Is the project then a failure or a success? This multilevel view is supported by Thomas and Fernández (2008), who focus not on five, but three levels, i.e. project management, technical, and business.

Sustainability

Sustainability is to manage the portfolio in such a way that it should be sustained indefinitely or in other words to implement the vision and strategies of the organisation without the depletion or destruction of resources.

Silvius, Ron Schipper, Planko, Van den Brink, & Köhler (2012) list three dimensions of sustainability that need to be incorporated within a portfolio. The dimensions are social, economic, and environmental. Harmony should be created between these dimensions (Figure 7.3).

The challenge that a portfolio manager faces is how to balance the three dimensions within a portfolio. There are principles that a portfolio manager can adopt to ensure the sustainability of the portfolio itself and ultimately that of the organisation.

The first principle is to think about all three dimensions. The organisation's focus is to make profits and create shareholder value. It is the duty of the portfolio manager to incorporate the social and environmental dimensions as well. Two of the four success criteria are affected by this thinking. The first is the success of the individual projects. Project success should also be measured based on sustainability focusing on the environmental

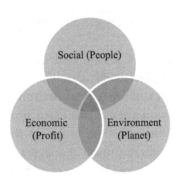

FIGURE 7.3
The three dimensions of sustainability.

and sociable dimensions. The second criterion is that of strategic success. Organisations cannot be successful in the long run if they are not sustainable.

The portfolio manager must think long term and focus on the benefits that each portfolio component will deliver. The benefits are normally reaped after the closure of a project. The benefits as stipulated in the business case should include economic, social, and environmental benefits. These benefits go beyond the quick wins of a successful project implementation.

Organisations are operating globally even if they are locally situated in a country or state. A portfolio might consist of portfolio components, which are implemented across the globe. The thinking of a portfolio manager should be locally and not globally. The advent of technology can assist a portfolio manager to have virtual meetings and therefore reduce the emission of gas as travelling is reduced. Care should be taken in the selection and balancing of a portfolio where the impact of portfolio components must be measured across all three dimensions.

One of the benefits of a Portfolio MIS is that resources are scheduled across all the portfolio components. A portfolio manager must remember that one of the principles of sustainability is that resources should not be depleted. This is especially the case when it comes to human resources. Resources should be managed in a sustainable manner where their performance on a current portfolio component should not have a negative impact on future performance.

Silvius et al. (2012) provide a basic checklist that can be used to incorporate sustainability into project management and ultimately into portfolio management. The checklist focuses on the three dimensions and allocates some aspects to these dimensions; e.g., under environmental sustainability one would get the notion of materials focusing on reusability and sourcing of said materials.

It is all fine and well to understand how to measure the success of a portfolio, but at some point in time that alone is not enough to ensure the continuous success of a portfolio. Portfolio managers must also understand that there are organisational factors that either contribute to the success of the portfolio or do not contribute.

These factors need to be embraced by the portfolio manager, and where possible, they should be optimised or increased to ensure the continuous success of the portfolio.

PORTFOLIO SUCCESS FACTORS

A portfolio success factor is a critical factor that is required to ensure the success of the portfolio. Although the absence or presence of a success criterion does not necessarily imply the automatic failure or success of a portfolio, it will certainly contribute to the success of the portfolio if all the success factors are addressed and followed (Figure 7.4).

Bolles & Hubbard (2007) identify various critical factors that might influence the successful implementation of a portfolio within an organisation. These factors include but are not limited to the following:

1. Total commitment and support from top executives: It is important that the executives of an organisation support portfolio management. If executives are not buying into the concept of portfolio management, how can we then expect that staff reporting to the executives will support the portfolio? Portfolio management spans across the entire organisation and touches the lives of every individual in the organisation. For this reason alone, executives should support the portfolio management. The executives can make commitments around financial and human resources from the start of portfolio management. Executives will also be able to provide the business value of portfolio management to the organisation.
2. The selection of a portfolio sponsor is critical. Although executives will support portfolio management, one individual is needed to drive and ensure the continuous support of portfolio management.

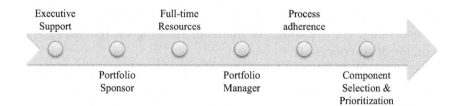

FIGURE 7.4
Portfolio success factors.

This individual, the portfolio sponsor, will be a strong propo-nent for portfolio management and will have some political cloud within the organisation. The sponsor will also serve as the liaison between the portfolio manager(s) and other primary decision-makers within the organisation.

3. Full-time resources: A portfolio cannot be managed on a part-time basis by someone who does it whenever there is time for it. Dedicated resources are needed to manage the portfolio on a day-to-day basis. These individuals are employees with specific skills, knowledge, and principles that are needed to ensure the successful management of the portfolio.

4. Following the preceding argumentation, the impact of a strong proj-ect portfolio manager is highly dependent on the influence of asso-ciated management roles. The direct and indirect influences of the senior management and the line management on the project port-folio management system that goes beyond the execution of certain project portfolio management (PPM) tasks will in the following be considered under the term of management involvement (Jonas, 2010).

5. Adherence to processes and procedures (governance): Policies and procedures need to be adhered to especially when it comes to the selection and prioritisation of portfolio components. The portfolio sponsor and respective portfolio managers must ensure that the policies and processes are followed. This is to ensure that everyone is treated fairly based on the policies and processes. If the policies and processes are not adhered to, the credibility of the entire portfolio management process is under scrutiny.

6. Müller et al. (2008) show the positive relationship between strategic success and portfolio selection and project portfolio performance.

A few other studies found project prioritisation as part of the portfolio management process to be a key success factor (e.g. Cooper, Edgett & Kleinschmidt, 1999; Elonen & Artto, 2003; Fricke & Shenbar, 2000). As it is no end on itself, successful project portfolio management needs to contribute to the overall business objectives (Meskendahl, 2010).

MATURITY OF PORTFOLIO MANAGEMENT

Organisations need repeatable and successful performance of their portfolio(s). This should be accompanied by continuous improvements of all the portfolio management processes. The more uniform and consistently portfolio management processes are applied, and the more mature these processes are, the greater are the positive results and realised benefits (Bolles & Hubbard, 2007). The benefit of mature portfolio processes is that the consistent application of these processes results in the delivery of portfolio components within the allocated budget and time frame. This will then eventually result in the continuous success of the portfolio and ultimately the success of the organisation.

The development and sustainment of mature portfolio management processes require the dedication, direction and involvement of the organisational executives and the portfolio sponsor.

Portfolio management should provide accurate, routine, consistent and adequate information into the progress of each of the portfolio components in order for the executives to take effective action when there is a deviation from the approved success criteria. Executives however have difficulty to understand the benefits of mature portfolio management processes for the following reasons:

1. Executives cannot match the results of mature processes to the financial results.
2. Executives do not understand how the attainment of portfolio management capabilities at a predefined maturity level relates to the cost of effective operations.
3. Executives cannot determine when the processes defining a maturity level would be finally standardised and become part of the organisation.

A four-stage portfolio management maturity model is presented for evaluating the progress made or not made regarding the implementation of portfolio management (Figure 7.5).

The very first step is to determine the baseline of the portfolio management processes. This is used to measure the progress of the portfolio management maturity levels. Stage 1 focuses on the allocation of additional resources and funds to develop the necessary processes to achieve stage 2. To move from stage 2 to stage 3, the newly developed processes, must be documented and implemented to achieve the desired results and benefits. Stage 3 establishes the processes as part of the day-to-day operations of the portfolio. The level of required resources drops to normal operational costs as no additional resources are needed to maintain the processes. Stage 4 is the desired state where all the portfolio management processes are optimised and improved upon in a continuous way.

AUDITING OF PORTFOLIO

Müller et al. (2008) states that previous studies on portfolio management did not address the notion of portfolio control as a factor influencing portfolio management performance. However, they all include supporting evidence that points towards examining this phenomenon further. The auditing of the portfolio can be used as a control factor.

Project auditing and therefore portfolio auditing originate from financial auditing, which is defined as "an independent, objective assurance and consulting activity designed to add value and improve an organisation's operations" (Institute of Internal Auditors (IIA), 2013). In the case of portfolio management, the focus of the audit should be to determine whether the various portfolio components contribute to the overall portfolio success and ultimately the success of the organisation. It must also be noted that the portfolio audit must be done by independent auditors. The Project Management Office (PMO) can play a role in this regard.

A portfolio audit will help an organisation to accomplish its objectives by bringing a systematic, disciplined approach to evaluate and improve the effectiveness of risk management, control, and governance processes (The Institute of Internal Auditors, 2013). By nature, auditing is a reactive process, and the main objective of portfolio auditing is the early detection of errors and deviations (Kumāra & Sharma, 2011).

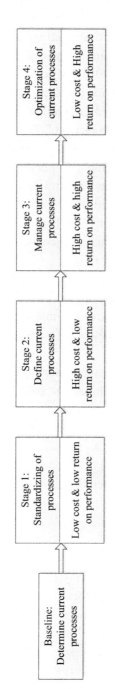

FIGURE 7.5

Four-stage portfolio management maturity model.

If the focus of attention moves to current portfolio and project management standards, then it is obvious that there is not really emphasis on the auditing of projects. Project audits are defined as the monitoring of compliance with project management standards, policies and procedures (Project Management Institute, 2013a). Another definition is that the purpose of a project audit is to provide an objective evaluation of the project (Association for Project Management, 2006). Neither PRINCE2 nor the P2M makes any reference to project auditing per se. The *Standard for Portfolio Management* does not make any reference to the auditing of a portfolio apart from the fact that the portfolio manager should have expertise and knowledge on the auditing of programmes and projects (Project Management Institute, 2013b).

If the advice of Gray and Larson (2008) as well as Hill (2007) is extrapolated to portfolio management, then the benefits of a portfolio audit should be as follows:

- The continuous monitoring of portfolio components' contributions to the achievement of the business objectives;
- The identification and response to weak and troubled portfolio performance;
- The overseeing of quality management activities;
- The maintaining of professional and best practices within the portfolio and its various components; and
- The compliance with organisational policies, government regulations and contractual obligations.

A distinction can be made between the various types of portfolio audits as discussed in the next section.

Types of Portfolio Audits

Figure 7.6 is a graphical presentation of the four major types of portfolio audits that an organisation can perform.

1. *Portfolio management audit*: This type of audit provides a comprehensive examination of the overall performance of portfolio management per se. The primary purpose is to ensure that the portfolio manager has put into place both business and technical processes that are likely to result in a successful portfolio.

FIGURE 7.6
Types of portfolio audits.

2. *Portfolio performance audit*: In contrast to the portfolio management audit, the portfolio performance audit represents a detailed examination of the financial and business aspects of the portfolio. Typical elements that will be audited are the four portfolio success criteria as per Figure 7.1.

3. *Portfolio management methodology*: Another important audit is that of the portfolio management methodology that is used in the organisation. This audit validates the content and effectiveness of the adopted portfolio management methodology. The highest ranked benefit of this type of auditing has been found to be client confidence, followed by enhanced accountability, reduced project costs and disputes in that order of significance (Sichombo, Muya, Shakantu, & Kaliba, 2009).

4. *Portfolio management processes*: This type of audit audits the process of portfolio management itself and the focus should include the following processes:

   a. Strategic management: The audit will focus on the processes that are followed to develop the strategy and roadmap of the portfolio and the subsequent alignment of these to the organisational strategy and objectives.

   b. Governance: The audit should focus on the processes of selecting and balancing the portfolio as discussed in Section "Selection and Balancing". Emphasis should be whether the processes are consistently applied and where there are deviations, these deviations are properly documented.

   c. Communication: The audit should determine whether the processes that are needed to develop the communication plan and the subsequent management of portfolio information are in place.

The communication processes must satisfy the information needs of all the stakeholders in order for effective and efficient decision-making with regard to the performance of the portfolio.

d. Risk: The audit should focus whether there is a structured process in place for the assessment and analysis of portfolio risks. The goal of portfolio risk management is to capitalise on potential opportunities and to mitigate those events or circumstances which can adversely impact the performance of the portfolio.

Portfolio audits, conducted routinely throughout the portfolio's lifespan, can help ensure organisational success through the identification of major risks that are likely to be faced by the portfolio manager, stimulating the portfolio manager and all the stakeholders to address the risks before it is too late to have a positive impact (McDonald, 2002). A portfolio audit is of no value to the organisation if the audit findings are not addressed. The purpose of the portfolio audit findings is to identify lessons learned that can help improve the performance of a project or of future projects (Stanleigh, 2009). The lessons learned can be applied to all the portfolio components as well as the portfolio itself.

Intervals of Auditing

A survey done by PWC (2012) stated that more than half of all respondents reported that they reviewed their portfolio on a monthly basis and 20% reported more frequent reviews. Respondents with monthly review cycles reported significantly higher rates of portfolio performance especially on business benefits. The opposite was also true from the survey where quarterly or less frequent portfolio reviews are associated with a decline in the benefits of portfolio management.

The importance of frequent audits is highlighted by this report from Pricewaterhouse Coopers (PWC). But it also places a huge burden on the organisation itself to perform regular portfolio audits. It is suggested that the following intervals are used for portfolio audits, i.e. monthly and quarterly.

1. The monthly audit can evolve around the overall performance of the portfolio. The focus should be on auditing of the methodology and processes that are followed within the portfolio. It will also focus on the performance of the individual portfolio components as per

Section "Portfolio Success Factors". The focus of the monthly audit is to ensure that everyone involved in the management of the portfolio follows the agreed-upon methodology and processes. This will in return improve the maturity of portfolio management as the audit will highlight deficiencies in the processes.

2. The quarterly audit can focus on the auditing of portfolio management itself and the overall performance of the portfolio. The audit will focus specifically on the success criteria of the portfolio as discussed in Section "Introduction". The reason for a quarterly audit is that changes to the financial value, strategic success and portfolio construction are not seen and experienced immediately. Portfolio components are added on an irregular basis, for instance every quarter or every six months.

It is important to highlight that these different types of audit activities require audit resources and should be part of the overall management of the portfolios. In fulfilling their duties, portfolio auditors must adhere to auditing standards, principles and the code of ethics defined by different standard-setting organisations and bodies to which they belong. It is also important that the auditors are independent to portfolio. "An independent [portfolio] audit committee fulfils a vital role in corporate governance. The [portfolio] audit committee is vital to ensure the integrity of integrated reporting and internal financial controls and identify and manage financial risks" (Institute of Directors Southern Africa, 2009:56). This statement implies that project audits fulfil a vital role in portfolio governance. Portfolio governance provides the structure through which the objectives of a portfolio are set, and the means of attaining those objectives and monitoring performance are determined (adapted from Turner, 2006). Portfolio audits, on the other hand, determine whether portfolio governance is in place, i.e. determining and attaining the portfolio objectives as well as monitoring the results of the objectives.

GOVERNANCE

Governance can be defined as a system or method of management (Bolles & Hubbard, 2007). Portfolio governance establishes the roles, responsibilities and authorities of each individual, the rules of conduct and management protocols. The establishment of portfolio structures within the organisation is needed as it will institute the planning and management processes of portfolio management. This view is supported by Mosavi (2014), who states that governance deals with roles and responsibilities, the decision-making frameworks, accountability, transparency, risk management, ethics, performance, and implementation of strategy.

Portfolio governance is the governance of the portfolio itself and focuses on the interrelationship between individuals, bodies, roles and responsibilities, decision-making processes and other governance elements at the portfolio level (Mosavi, 2014).

Portfolio Management Roles and Responsibilities

Organisations must create the functional structures that own and provide portfolio management. These structures are needed for the adequate and effective planning, authorisation and chartering of portfolios. The benefits of having these functional structures in place are the following:

1. The assurance that all the portfolio components are aligned with the organisational strategy and objectives.
2. Integration is provided for the monitoring and controlling of portfolio components.
3. Ownership and maintenance of portfolio processes, procedures and templates.
4. Identification and categorisation of each business objective into a specific portfolio.

The following portfolio roles are necessary for the effective management of the portfolio:

Portfolio Steering Committee

One structure that needs to be in place is that of the portfolio steering committee. Portfolio steering committees have the delegated authority to govern the portfolio but need to focus on the following aspects:

1. The portfolio steering committee must communicate and confirm the decisions that are made. These decisions might include the selection and optimisation of portfolio components, the general status of the portfolio, and the release of the various audit reports.

2. The portfolio steering committee also performs the role of negotiators. The negotiation is to determine whether a new project should be part of the portfolio and secondly when will it become part of the portfolio.
3. The portfolio steering committee must also make timely decisions. It is up to the portfolio steering committee to make decisions that affect the overall performance of the portfolio, and these decisions must be made in a quick and decisive manner.

There are also other factors that play a role such as the frequency and duration of the portfolio steering committee meetings. These meetings must take place in accordance with the mandate. Depending on the complexity of the portfolio but also the maturity levels, meetings can be held monthly, quarterly or every six months. The portfolio steering committee must take into consideration the various audit reports that need to be discussed as well as the selection and optimisation of the portfolio components when it decides the frequency of the meetings. Research has established the most portfolio steering committees are spending half a day to a full day on portfolio-related issues.

The portfolio steering committee normally consists of portfolio managers, project owners, and representatives from the top management. One important aspect is that there must be trust and open communication between the various members of the portfolio steering committee.

Project Owners
Project owners are typically middle managers who are responsible for the implementation of projects and programmes. According to Blomquist & Müller (2006), these project owners are not formally part of portfolio management but are co-opted based on the state of the project. Project owners engage with the portfolio before and during the execution of the project. The role of the project owner is to identify non-performing projects, participate in the portfolio steering group, and perform administrative tasks relate to the management of a project. The aim is to improve the efficiency and success rates of a project, which is a criterion for a successful portfolio.

Top Managers
Top managers or someone representing them should also form part of the portfolio steering committee. Their role is to ensure that everyone understands the organisational vision and strategies. The purpose of portfolio management is to ensure that the strategies are successfully implemented.

This is not possible if there is not a clear understanding of the organisation's vision and strategies. Top managers will ensure that the vision and strategies are foremost at any meeting. They will also inform the portfolio steering committee about changes to the vision and strategies of the organisation. Such changes will have an impact on the portfolio itself. Portfolio components will have to be re-evaluated, and new components might be selected in favour of current portfolio components.

Portfolio Manager

This role is assigned to one individual only. The portfolio manager analyses all future and current portfolio components and recommends a viable mix. The portfolio manager will monitor the performance of the portfolio on a daily basis and report to the steering committee if there are any deviations. The portfolio manager also plays an important role in supporting the overall strategy of the organisation and prepares information for the steering committee.

Organisations must also realise that the governance structures can and must be adapted to suit the needs of the organisation. There is no silver bullet telling organisations what the governance structures are and how they should operate.

Portfolio Management Office (PfMO)

The role of the PfMO within the realm of portfolio management is to facilitate the involvement of the executives and senior managers in the oversight of portfolio components (Hill, 2007). The PfMO can develop and implement processes and procedures for each of the portfolio management activities, and this should be consistent with the maturity levels. Hill (2007) and the Project Management Institute (2013b) state that the following activities should be performed by the PfMO:

1. Development of the process and criteria for project selection
2. Collection of portfolio component performance data for comparison purposes
3. Alignment of portfolio components with the organisational strategy
4. Establishment of gate reviews that are in line with the auditing of the portfolio
5. Implementation of a process to review the availability of resources
6. The use of real-time data for decision-making

In spite of all these activities that a PfMO must fulfil, Unger, Gemünden, & Aubry (2012) state that the PfMO must fit into the overall organisational management framework. They continue to state that the PfMO must have a strong governance focus that ensures the implementation of the organisational goals. This governance aspect of the PfMO focuses on three tiers:

1. Coordinating, which includes the facilitation of cross-portfolio components and the allocation of resources
2. Controlling, focusing on the establishment and updating of information for accurate decision-making and including advice on corrective measures to the portfolio steering committee
3. Supporting, which provides services as per Hill (2007) to the various managers of the portfolio components and ultimately the cultivation of portfolio management standards and procedures

The PfMO must first of all perform the three governance roles and then align the respective roles to these roles. By doing this, it is assured that the PfMO performs its duties in accordance with the strategies of the organisation.

Portfolio Management Information System

A portfolio manager needs to understand what progress is made within the portfolio with regard to the success criteria, and this needs to be done fairly frequently to minimise risk exposure and to maximise opportunities (Kendall & Rollins, 2003).

Access to real-time portfolio information is also a key factor in the successful management of a portfolio (Rajegopal, McGuin, & Waller, 2007). Real-time information is needed to make decisions on unforeseen business changes as well as unexpected opportunities within the environment. The purpose of a portfolio Management Information System (MIS) is to get relevant information into the hands of decision-makers as soon as possible.

Information availability for decision-makers is shown as the most significant project-level factor contributing to portfolio management efficiency both directly and through project management efficiency (Müller et al., 2008). The following are the basic functionalities that a portfolio MIS should provide to the organisation:

- Define and establish a common communication and reporting platform for all projects in the portfolio. Measure and compare projects along similar metrics.

- Take portfolio decisions in teams, after evaluation of the pros and cons of different mixes of priorities and go/no-go decisions, at the organisational and the portfolio level.
- Establish organisational routines to ensure project selection based on the organisation's strategy, not the personal preferences of individual managers.
- Review portfolios periodically using comparable metrics.

There are various role players in the industry that provide portfolio MIS solutions. Some of the organisations include Primavera (http://www.oracle.com/us/products/applications/primavera/p6-enterprise-project-portfolio-management/overview/index.html), Sciforma (http://www.sciforma.com/en-uk/) and Microsoft (http://office.microsoft.com/en-za/project/project-portfolio-management-for-the-enterprise-project-server-FX103802061.aspx). Figure 7.7 is a presentation from a Primavera screenshot illustrating how a portfolio is presented.

All portfolio MIS solutions should provide the following functionality:

- Assist in the planning, scheduling, and controlling of large-scale programs and individual projects
- Assist in the selection of the right strategic mix of projects
- Balance resource capacity
- Assist in the allocation of the best resources and track progress
- Monitor and visualise project performance versus planned performance
- Foster team collaboration
- Integrate with financial management and human capital management systems
- Simulate the impacts of a project on the entire portfolio
- Draw up a long-term project plan, using "what if" scenarios
- Breakdown the relevant sections of the long-term plan into operational projects
- Update the portfolio according to actuals
- Achieve a balanced scorecard by meeting the organisation's vision and strategies while managing the portfolios and the portfolio components

Organisations should use both the control mechanisms and the success factors and assign operational values to them. That allows tracking,

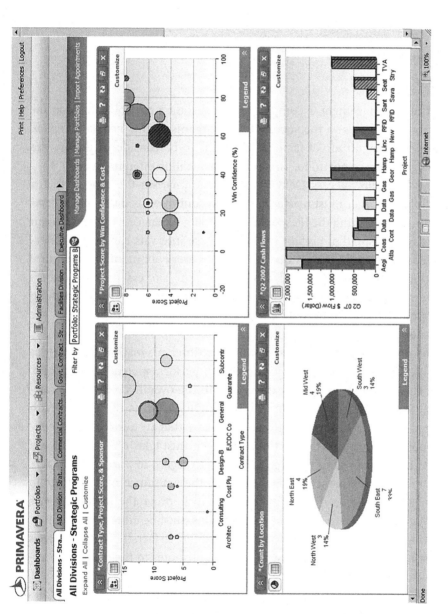

FIGURE 7.7
Primavera portfolio dashboard.

setting, and measurement of goals, and their achievement, which is the aim of portfolio management. The MIS systems can highly successful be used to achieve this objective.

CONCLUSION

Portfolio management is complex and to sustain a portfolio that continuously contributes to the success of the organisation is just as complex. There is a magnitude of factors that needs to be considered during the management of a portfolio. This complexity is illustrated in Figure 7.8.

The Portfolio Success Delivery Model summarises the aspects that a portfolio manager must master in order to continuously deliver a successful portfolio. During the lifespan of a portfolio, the success of the portfolio must be measured at certain intervals. These intervals can be pre-determined for example once a quarter or once every six months. The entire portfolio must be measured against the four success criteria, under-performing portfolio components can be removed, and new portfolio components can be added based on the vision and strategies of the organisation.

The portfolio manager must also at these intervals institute the various audits. The audits can assist in the overall performance of the portfolio. The audits will highlight any deviations to the overall vision and strategies. The results of the audit will also highlight areas of concern with regard to the portfolio management processes that are adhered to or not.

Another aspect that the portfolio manager must take into consideration is the maturity level of the organisation with regard to portfolio management. This assessment will be done with the assistance of the Portfolio Management Office. The ultimate goal is to improve the maturity levels throughout the lifespan of the portfolio. This must be done without interrupting the overall performance of the portfolio. In fact, the maturity levels must be improved in such a way that it contributes to the overall performance of the portfolio.

The success of the portfolio cannot be guaranteed with certain factors in place. A few factors have been discussed that can have a positive impact on the success of the portfolio. The portfolio manager must continuously engage with the various stakeholders and sponsors to ensure that the political will is there to support the portfolio even when some

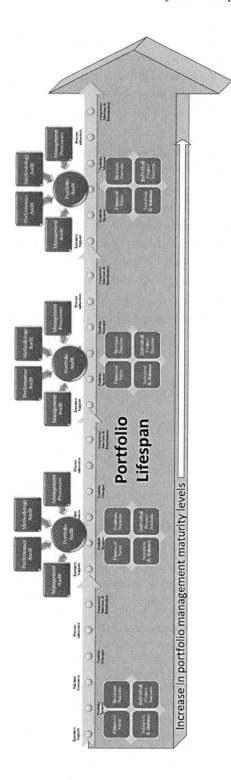

FIGURE 7.8
Portfolio success delivery model.

decisions taken by the portfolio manager is not popular. The portfolio manager must also observe the environment to determine whether there are new or different factors that contribute to the overall success of the portfolio. It is also of the essence to look at factors that do not contribute to the overall success of the portfolio and minimise those factors impact.

All this balancing and cross-checking must be done within the constraints of sustainability and governance. It does not make sense to drive the success agenda of the portfolio if it is not sustainable in the long run. Any decision made by the portfolio manager must be weighed against the three dimension of sustainability. None of the three dimensions are more important that any of the other dimensions. In that lies the difficulty for the portfolio manager as the economic dimension will always get preference.

The role of a portfolio manager is complex, and managing a portfolio in a successful way and manner is daunting. The Portfolio Success Delivery Model simplifies the complexity and highlights the aspects that the portfolio manager must focus on continually. The challenge for the portfolio manager is to ensure that all the aspects are taken into consideration during the lifespan of the portfolio.

REFERENCES

Association for Project Management. (2006). *APM Body of Knowledge* (5th ed.). Buckinghamshire: Association for Project Management.

Bannerman, P. L. (2008). *Defining project success: a multilevel framework.* Paper presented at the *PMI Research Conference: Defining the future of project management*, Warsaw, Poland.

Bennington, P., & Baccarini, D. (2004). Project benefits management in IT projects – An Australian perspective. *Project Management Journal, 35*(2), 20–31.

Beringer, C., Jonas, D., & Kock, A. (2013). Behavior of internal stakeholders in project portfolio management and its impact on success. *International Journal of Project Management, 31*(6), 830–846. doi: 10.1016/j.ijproman.2012.11.006

Blomquist, T., & Müller, R. (2006). Practices, roles, and responsibilities of middle managers in program and portfolio management. *Project Management Journal, 37*(1), 52–66.

Bolles, D. L., & Hubbard, D. G. (2007). *The Power of Enterprise-wide Project Management.* New York: American Management Association.

Bradley, B. (2010). *Benefit Realisation Management: A Practical Guide to Achieving Benefits Through Change* (2nd ed.). Surrey, UK: Gower Publishing Limited.

Chen, C.-T., & Cheng, H.-L. (2009). A comprehensive model for selecting information system projects under fuzzy environment. *International Journal of Project Management, 27*(4), 389–399.

Cooke-Davies, T. J. (2005). *The executive sponsor – The hinge upon which organisational project management maturity turns.* Paper presented at the *PMI Global Congress*, Edinburgh, Scotland.

Cooper, R., Edgett, S., & Kleinschmidt, E. (1999). New product portfolio management: Practices and performance. *Journal of Product Innovation Management, 16*, 333–351.

De Reyck, B., Grushka-Cockayne, Y., Lockett, M., Calderini, S. R., Moura, M., & Sloper, A. (2005). The impact of project portfolio management on information technology projects. *International Journal of Project Management, 23*(7), 524–537. doi: 10.1016/j.ijproman.2005.02.003

Dekkers, C., & Forselius, P. (2007). *Increase ICT project success with concrete scope management.* Paper presented at the 33rd EUROMICRO Conference on Software Engineering and Advanced Applications, 2007.

Dhillon, G. (2005). Gaining benefits from IS/IT implementation: Interpretations from case studies. *International Journal of Information Management, 25*(6), 502–515.

Eckartz, S., Daneva, M., Wieringa, R., & Hillegersberg, J. V. (2009). *Cross-organizational ERP management: how to create a successful business case?* Paper presented at the *Proceedings of the 2009 ACM symposium on Applied Computing*, Honolulu, Hawaii.

Elonen, S., & Artto, K. A. (2003). Problems in managing internal development projects in multi-project environments. *International Journal of Project Management, 21*(6), 395–402. doi: 10.1016/S0263-7863(02)00097-2

Fricke, S. E., & Shenbar, A. J. (2000). Managing multiple engineering projects in a manufacturing support environment. *IEEE Transactions on Engineering Management, 47*(2), 258–268. doi:10.1109/17.846792

Gray, C. F., & Larson, E. W. (2008). *Project Management: The Managerial Process* (4thv). New York: McGraw-Hill/Irwin.

Hill, G. M. (2007). *The Complete Project Management Office Handbook* (2nd ed.). Boca Raton, FL: Auerbach Publications.

Hyväri, I. (2006). Success of projects in different organizational conditions. *Project Management Journal, 37*(4), 31–41.

Institute of Directors Southern Africa. (2009). *King Code of Governance for South Africa 2009.* Johannesburg: Institute of Directors Southern Africa.

Institute of Internal Auditors. (2013). *Definition of Internal Auditing.* Available from http://www.theiia.org/guidance/standards-and-guidance/ippf/definition-of-internal-auditing/?search%C2%BCdefinition. Accessed 31 March 2014.

International Institute of Business Analysis. (2009). *A Guide to the Business Analysis Body of Knowledge (BABOK Guide)* (2nd ed.). Toronto, Canada: International Institute of Business Analysis.

Jonas, D. (2010). Empowering project portfolio managers: How management involvement impacts project portfolio management performance. *International Journal of Project Management, 28*(8), 818–831. doi: 10.1016/j.ijproman.2010.07.002

Kaplan, R. S., & Norton, D. P. (2004). *Strategy Maps: Converting Intangible Assets into Tangible Outcomes.* Boston, MA: Harvard Business School Press.

Kendall, G., & Rollins, S. (2003). *Advanced Project Portfolio Management and the PMO:* Boca Raton, FL: J. Ross Publishing.

Kumāra, R., & Sharma, V. (2011). *Auditing: Principles and Practice* (2nd ed.). New Delhi, India: PHI Learning.

Marnewick, C. (2012). *A longitudinal analysis of ICT project success.* Paper presented at the *Proceedings of the South African Institute for Computer Scientists and Information Technologists Conference*, Pretoria, South Africa.

McDonald, J. (2002). Software project management audits – update and experience report. *Journal of Systems and Software, 64*(3), 247–255.

Meskendahl, S. (2010). The influence of business strategy on project portfolio management and its success – A conceptual framework. *International Journal of Project Management, 28*(8), 807–817. doi: 10.1016/j.ijproman.2010.06.007

Mosavi, A. (2014). Exploring the roles of portfolio steering committees in project portfolio governance. *International Journal of Project Management, 32*(3), 388–399. doi: 10.1016/j.ijproman.2013.07.004

Müller, R., Martinsuo, M., & Blomquist, T. (2008). Project portfolio control and portfolio management performance in different contexts. *Project Management Journal, 39*(3), 28–42. doi: 10.1002/pmj.20053

Office of Government Commerce. (2003). *Managing Successful Projects with PRINCE2*. London: Office of Government Commerce.

Ohara, S. (2005). *P2M: A Guidebook of Project & Program Management for Enterprise Innovation* (3rd ed.): Tokyo, Japan: Project Management Association of Japan.

Project Management Institute. (2008). *A Guide to the Project Management Body of Knowledge (PMBOK Guide)* (4th ed.). Newtown Square, PA: Project Management Institute.

Project Management Institute. (2013a). *A Guide to the Project Management Body of Knowledge (PMBOK Guide)* (5th ed.). Newtown Square, PA: Project Management Institute.

Project Management Institute. (2013b). *The Standard for Portfolio Management* (3rd ed.). Newtown Square, PA: Project Management Institute.

PWC. (2012). *Insights and Trends: Current Portfolio, Programme, and Project Management Practices* (pp. 40): London, UK: PWC.

Rajegopal, S., McGuin, P., & Waller, J. (2007). *Project Portfolio Management: Leading the Corporate Vision*. New York: Palgrave Macmillan.

Reilly, F. K., & Brown, K. C. (2012). *Analysis of Investments and Management of Portfolios*. United States of America: Cengage Learning.

Remenyi, D., Money, A., & Bannister, F. (2007). *The Effective Measurement and Management of ICT Costs and Benefits* (3rd ed.). Oxford, UK: CIMA Publishing.

Remenyi, D., & Sherwood-Smith, M. (1998). Business benefits from information systems through an active benefits realisation programme. *International Journal of Project Management, 16*(2), 81–98.

Sichombo, B., Muya, M., Shakantu, W., & Kaliba, C. (2009). The need for technical auditing in the Zambian construction industry. *International Journal of Project Management, 27*(8), 821–832.

Silvius, G., Ron Schipper, R., Planko, J., Van den Brink, J., & Köhler, A. (2012). *Sustainability in Project Management*. Surrey, UK: Gower Publishing.

Stanleigh, M. (2009). Undertaking a successful project audit: Business Improvement Architects. Available from http://www.projectsmart.co.uk/undertaking-a-successful-project-audit.html. Accessed 31 March 2014.

Stewart, R. A. (2008). A framework for the life cycle management of information technology projects: ProjectIT. *International Journal of Project Management, 26*(2), 203–212.

Thomas, G., & Fernández, W. (2008). Success in IT projects: A matter of definition? *International Journal of Project Management, 26*(7), 733–742.

Turner, J. R. (2006). Towards a theory of project management: The nature of the project governance and project management. *International Journal of Project Management, 24*(2), 93–95.

Unger, B. N., Gemünden, H. G., & Aubry, M. (2012). The three roles of a project portfolio management office: Their impact on portfolio management execution and success. *International Journal of Project Management, 30*(5), 608–620. doi: 10.1016/j.ijproman.2012.01.015

Williams, D., & Parr, T. (2008) *Enterprise Programme Management – Delivering Value* (pp. 303). New York: Palgrave MacMillan.

8

Strategic Portfolio Management through Effective Communications

Amaury Aubrée-Dauchez

INTRODUCTION

Challenging economic, financial, and social conditions force organizations to gain competitive hedge optimizing operations and to implement change. Such change must be orchestrated through portfolio management to ensure its high alignment to business strategy, mission, goals, and objectives; it must genuinely put emphasis on effective governance and compliance while leveraging best practices and platforms and managing risks. However, to get value for money, organizations must place communication on the forefront enabling controlled, secured, and seamless flows with whomever is concerned, whether collaboration relates to normal work or project portfolio, whether communication is internal or involves external entities such as partners, vendors, customers, authorities, bloggers, journalists, etc.

Amid recent global pandemic as well as current and forthcoming deep economic woes, the organizations communicating better, both internally and externally, have proven being more resilient. Among organizations that are considered highly effective communicators, 80% of projects meet original goals, versus only 52% at their minimally effective counterparts, according to PMI's *Pulse of the Profession In-Depth Report: The Essential Role of Communications* (2013b). The same report – still very valid in its conclusions and the order of magnitude – revealed that US

DOI: 10.1201/9781003315902-8

$135 million is at risk for every US $1 billion invested in the project portfolio. Further research on the importance of effective communications uncovers that 56% (US $75 million of that US $135 million) is at risk because of ineffective communications.

This chapter defines the concepts involved and how portfolio management and communication, including collaboration aspects, have evolved during the last decades both from a process and an enabler standpoint including a review of some of the literature in the field. Next, it describes the importance of communication to strategic portfolio management. It is followed by a discussion to present how communication should be implemented around strategic portfolio management in order to maximize its impact and relevance throughout the organization and the key considerations during the implementation process. A summary concludes the chapter.

EVOLVING CONCEPTS AND PROCESSES

Portfolio Management Becomes More Strategic

There are many factors impacting the rise of portfolio management as a business discipline including the propagation of Markowitz's portfolio theory (1952), a growing focus on cost visibility following various financial crises, corporate scandals, and project failures, and the importance given organization wide to accountability and governance relying on compliance and standardization mechanisms. Portfolio management attempts to answer the "Catch 22" problem according to which:

> Strategy without action is only a daydream, but action without strategy is a nightmare.

As a matter of fact, strategy must have a vision which implies change; changes are orchestrated through ad hoc works, projects, programs, and sub-portfolios. However, experience shows (Royer, 2003) that, whether or not they deliver on their promises, these activities tend to justify their own existence if they do not loop back to the strategy. While project and program management focus on "doing the work right", the purpose of portfolio management is "doing the right work".

While the first edition (2006) of the *Standard for Portfolio Management* (Project Management Institute, 2006) defines portfolio management as "an approach to achieving strategic goals by selecting, prioritizing, assessing, and managing projects, programs, even portfolios as well as other related work based upon their alignment and contribution to the organization's strategies and objectives", the third edition (2012) simplified it to "the centralized management of one or more portfolios to achieve strategic objectives".

Since inception, portfolio management is an integral part of the organization's overall strategic plan. This being said, for many years, the discipline was rather perceived just as information technology (IT) portfolio management with outputs and benefits belonging to the support side of the organization and the Chief Information Officer (CIO) responsible for the oversight. More recently, other members of the Chief Executive Officer's team (CxOs) have started to take the lead on portfolio steering. First, change is daily on company agendas; it has a bigger stake than before, and corporate mistakes have a much lower level of acceptance. Second, technology itself is changing the nature of the relationships between employees, managers, consumers, and partners with traditional border lines getting fuzzier: operations and support, internal and external, clients, and shareholders. Last, methodologies and supporting platforms are more extensive at bridging the gap between strategy and operational execution.

As outlined by Levin, Artl, and Ward (2010) in practice, however, few organizations have a complete understanding of all the work that is under way in the organization, even if an enterprise-wide portfolio management office is in place. While the Chief Executive Officer (CEO) and Chief Portfolio Officer (CPO) focus on the large programs and complex projects, they probably are not aware of many of the smaller projects that are in progress and do not follow the portfolio management process. What is important to understand is that strategic portfolio management communication involves third parties. Communication is instrumental in uniting people, organizational units, vendors, and partners to overarching mission, goals, and objectives.

Communication: From Linearity to Simultaneity

The first major model for communication was introduced by Shannon and Weaver (1949) for Bell Laboratories. The original model was designed to mirror the functioning of radio and telephone technologies and consisted of three primary parts: sender, channel, and receiver. In that model, communication is the process by which a person, group, or organization (the sender) transmits some type of information (the message) to another person, group, or organization (the receiver).

In later extensions of such linear model (Berlo, 1960), communication becomes social interaction where at least two interacting agents share a common set of signs and rules. In that respect, collaboration, i.e. working with each other to do tasks and to achieve shared goals (Collins English Dictionary, 2012), becomes tightly imbricated with communication. Indeed, when collaboration goes across organizational fences it is not only a means to optimize resources and share knowledge among portfolios, programs, and projects, it is also one of the most important strategies applied to alleviate conflicts; when parties in conflict each desire to fully satisfy the concerns of all parties, there is cooperation and a search for a mutually beneficial outcome. In collaborating, the parties intend to solve problems by clarifying differences rather than by accommodating various points of view.

Beside formal and informal lateral communication, the portfolio management ecosystem combines upward and downward communication. Robbins & Judge (2013) define the former as a flow to a higher level in the group or organization used to provide feedback to higher-ups, inform them of progress toward goals, and relay current problems, whereas the latter is a flow from one level of a group or organization to a lower level used to assign goals, provide job instructions, explain policies and procedures, point out problems that need attention, and offer feedback about performance. However, despite its wide acceptance derived from its simplicity, generality, and quantifiability, these approaches had a fundamental limitation: they assume that communicators are isolated individuals.

In light of these weaknesses, Barnlund (2008) proposed a transactional model of communication. The basic premise of the transactional model of communication is that individuals are simultaneously engaging in the sending and receiving of messages. Effective management focuses on ever-changing agendas of strategic issues. These agendas consist of multiple challenges, stretching aspirations, and ambiguous issues. When managers deal with the issues on their strategic agendas, they are performing a real-time learning activity.

There is no overall framework to which they can refer before they decide how to tackle an issue. Through discussion with each other and with customers, suppliers, and even competitors, they are discovering what objectives they should pursue and what actions might work. The communication is spontaneous in the sense that it is not directed by some central authority. The communication that occurs depends upon the individuals involved. And that depends upon the boundary conditions, or context, provided by individual personalities, the dynamics of their interaction with each other, and the time they have available, given all the other issues requiring attention.

While still operating within separated parts, communication and collaboration are now holistic and span across interconnected systems. Effective portfolio management must find an ever-moving balance between structured and unstructured flows of information. Brafman and Beckstrom (2008) explain that the decentralized "sweet spot" is the point along the centralized–decentralized continuum that yields the best competitive position. The organization must enable enough decentralization for creativity but requires sufficient structure and controls to ensure consistency and compliance. The forces of centralization and decentralization continue to pull the sweet spot to and from one another. In any industry that is based on information – whether it is music, software, or banking – these forces pull the sweet spot toward decentralization. The more security and accountability become the norm the more likely it is that the sweet spot will tend toward centralization.

Embedded Complexity

The evolution of portfolio management as well as communication as described in this chapter actually reflects changes both in our society and organizations. For executives, managers, employees, and partners, it is sometimes even perceived as a revolution. First, formal communication flows still exist; they are basically formally prescribed patterns of interrelationships between various elements of the organization often dictating who may and may not communicate with whom; however, they coexist with informal exchanges. Then, Davis and Meyer (1999) coined the term "blur" to describe the combined effect of speed, connectivity, and intangibles on the business world, speed being how fast change occurs, connectivity is how networked businesses are linked to one another to complete a

144 • Portfolio Management

value chain, and intangibles are the ratio of non-physical to physical assets. On that basis, Evans and Roth (2004) developed the notion of collaborative knowledge networks, which underline several topics addressed hereunder.

Through cross-cultural differences, diversity, and gender mainstreaming as well as new work practices, globalization, and social shifts contribute to increase to an even-higher portfolio management complexity:

- Cross-cultural difference. Effective communication is difficult under the best of conditions. Cross-cultural factors clearly create the potential for increased communication and collaboration problems. A gesture that is well understood and acceptable in one culture can be meaningless or lewd in another. Only a few companies have documented strategies for communicating with employees across cultures, and not many more require that corporate messages be customized for consumption in other cultures.
- Diversity and gender mainstreaming are the public policy concept of assessing the different implications for women and men of any planned policy action, including legislation and programs, in all areas and levels. While linked gender balance with mainstreaming essentially offers a pluralistic approach that values the diversity among both women and men, some organizations tackle these changes by making bold individual appointments to strengthen their commitment to this theme (Burston, 2012).
- New work practices. Shifts in company structure lead toward greater and freer communication and sharing of information, outsourcing, off-shoring, global teams, mobile workforce, flexible work arrangements, less hierarchical organizations, and the influx of the Millennial generation. About remote work, while not all companies allow employees to work off site, data from Gallup's State of the American Workplace report (2017) show that more than four in 10 (43%) of the employees surveyed spend some amount of time working remotely or in locations apart from their coworkers. This percentage can only have increased during and after the global pandemic with many organizations deploying platforms in a matter of week if not days to cope with imposed confinements. And, Gallup (2017) finds that companies that offer the opportunity to work remotely might have some advantages when it comes to hours worked and employee engagement.

EVOLVING SUPPORTING PLATFORMS

Background

In the early days, project management software ran on big mainframe computers and was used only in large projects. These early systems were limited in their capabilities and, by today's standards, were difficult to use. In May 1957, the Remington Rand Corporation and the DuPont Corporation started a joint venture to develop the Critical Path Method (CPM) mathematical technique for managing plant maintenance projects which allowed DuPont to save 25% on its plant shutdowns. This technique was popular but expensive and was later dropped by DuPont after a management change took place; it lost traction but came back even overtaking the Program Evaluation Review Technique (PERT) developed by the U.S. Navy until the "Precedence" technique used by many scheduling systems was developed in the 1970s. Nevertheless, according to Rich (2011), project management became the first commercial software program, which ran on a computer using a "stored program".

As Aubrée-Dauchez states (2005), in order to implement effective communications and enable proper collaboration among stakeholders, Project Portfolio Management (PPM) platforms shall provide much more than dashboards and monitoring reports. A discussion on supporting tools must be integrated into a broader context including, inter alia, the exchange of information between different stakeholders, document management systems, the different methodologies that can inspire or directly influence the activities or platform coding, and last but not least, the mandatory commitment of senior management. In this review, the technical layer as such is less important than the application layer (LEADing Practice, 2014); whether portfolio management users access information via a file server, the Intranet or the Internet, it does not affect the business logic per se.

Bottom-Up

With the acceleration of micro-computing in the 1970s and 1980s, organizations started to manage projects on personal computers with individuals or groups gathering in separate files summary information on progress, risks, and issues. This approach is sometimes referred to

Enterprise Portfolio Management (EPM) since it is geared to supporting day-to-day PPM.

This bottom-up solution also accommodates spreadsheets, text files, and even sometimes hand-written notes that have to be consolidated at the program and portfolio management levels. Such an approach proved to be cumbersome and prone to errors hindering collaboration and communication and making portfolio management difficult and unable to perform well in achieving strategic goals.

Top-Down

As noted by Levin et al. (2010), the emphasis therefore has changed to ensure that only those programs and projects that support the organization's strategic objectives are selected and pursued. Therefore, alongside of formalization and standardization of portfolio management practice, platforms started to emerge in the late 1990s to provide a top-down approach to managing, tracking, and reporting on enterprise strategies, projects, portfolios, processes, resources, and results.

This approach, however, faces inherent limitations; most top-down solutions tools fail to accommodate work other than projects such as unstructured work or ad hoc requests. Moreover, collaborative project elements and lower-level communication are not noise anymore in a business environment embedded in complexity. There is nothing more outdated than a static spreadsheet, even received a few minutes ago. With that in mind, most of the top-down solutions offer, at cost, integration connectors to synchronize, to the extent possible, data with other applications.

Integrated

In order to overcome the shortfalls of the unidirectional solutions and sometimes in parallel to the development of top-down solutions, integrated platforms have been made available on the market around year 2000. Basically, project information is stored in a central database and protected from unauthorized access and corruption. Project managers can drill down into project details, while portfolio managers can apply business logic to select, prioritize, and assess projects working dynamically on the same data sets and accessing the same document and discussion repositories. The integrated solution has organization-wide default and custom fields allowing comparison across the project portfolio.

About communication, it is worth noting that Aberdeen (2013) has termed Integrated Communications (IC) as the approach that enables organizations to integrate solutions from a variety of vendors. Unified Communications has long appeared as the mythic "Holy Grail" of business communications, the seamless integration of all communication channels: voice, voice messaging, email, Short Message Service (SMS), text, chat, or instant messaging (IM), presence, video, and fax with the ability to move a conversation effortlessly from one channel to the next. This promise has yet to be fulfilled for many organizations, in part because of a lack of a compelling business case, and until recently a lack of industry standards that would make agile, cross-channel business conversations viable. The relatively recent emergence of standards has driven down the cost and increased the adoption of advanced communications capabilities.

Social and Agile

Organizations often try to address business challenges by relying on the traditional managerial tools to pursue operational excellence: establishing well-defined roles, best practice processes, and formal accountability structures. However, Cross's research (2010) shows that such tools, though valuable, are not enough. The key to delivering both operational excellence and innovation is having networks of informal collaboration. Within departments of large global companies, Cross notes innovative solutions often emerge unexpectedly through informal and unplanned interactions between individuals who see problems from different perspectives. Additionally, successful execution frequently flows from the networks of relationships that help employees handle situations that do not fit cleanly into established processes and structures.

Organizational leaders, who learn to harness and balance both formal and informal structures, can create global organizations that are more efficient and innovative than organizations that rely primarily on formal mechanisms. However, even though individual employees may be able to identify local patterns of collaboration, broader configurations of informal collaboration tend to be far less visible to senior leaders. In the face of this reality, Cross found that organizational network analysis offers a useful methodology to help executives do two things: assess broader patterns of informal networks among individuals, teams, functions, and organizations and then take targeted steps to align networks with strategic imperatives.

Some PPM platforms acknowledge that communication and collaboration are much more than meeting and talking about work; it is about connecting workers to each other and to information whenever they need it, wherever they are. They have embraced Mobile First principles (Wroblewski, 2011), and social network user experience, they provide agile project management features out of the box, and they extend portfolio management to an end-to-end organization work life cycle. Forrester (2010) states that social networking software that enables people to post their activities and receive feedback from their peers has taken off like a rocket, particularly with consumers. This type of peer collaboration, exemplified in consumer-oriented social networks, represents the new design metaphor needed to make PPM more broadly accepted and productively used.

The management of the pandemic and the legal constraints have further reinforced these integration, social, and agile imperatives. The situation between 2020 and early 2022 was very problematic. The pandemic has forced organizations to close offices and to develop remote working. Physical distance brought new challenges: how can teams work together and how can organizations maintain or increase productivity when they are not physically present. Collaboration and communications tools such as Slack and Teams responded, in part, through chat and information sharing, but there are also a number of needs for better organizational capacities to monitor activities and automate processes and thus limit delays caused by the absence of direct contacts.

IMPORTANCE OF COMMUNICATION TO STRATEGIC PORTFOLIO MANAGEMENT

According to Locker (2001), experts consider communication to be a key process underlying all aspects of organizational management. Contemporary scholars have referred to organizational communication as "the social glue… that continues to keep the organization tied together" (Roberts, 1984) and "the essence of organization" (Weick, 1987). Writing many years earlier, Chester (1938, p. 91) said, "The structure, extensiveness and scope of the organization are almost entirely determined by communication techniques".

Since portfolio management provides key capabilities for achievement of an organization's strategy, there may be a major focus at the executive level both to assemble and to communicate detailed information on the progress of major objectives and impacts of the components, as well as any changes to previously communicated plans.

Ongoing and well-targeted communication is a key requirement for maintaining stakeholder confidence in and support for the objectives to be achieved and the approaches being implemented. In addition, in order to ensure coordination and effective teamwork, communication among the teams responsible for the various components need to planned, formalized, and managed in concert with best practices. Various portfolio events or milestones need to be communicated both inside and outside the organization. This could include achievement of a major objective, elimination of a component, and other matters requiring corporate communications.

The challenges are that manual project requests through different tools make it difficult to stay on top of opportunities, squeaky wheels demand attention, but distract from more important priorities, and manually collecting the status of a portfolio of projects is like "herding cats". The organization's leaders must validate, align, and optimize the right portfolio investments and communicate about them.

The strategic portfolio management system put in place must ensure that every project aligns with organization's goals and objectives by standardizing project requests; quickly validate, align, and prioritize projects in portfolios; and make the right decisions on trusted data. Gartner (2013) reports that PPM will evolve to cover the project life cycle from a bright idea to a post-project review whereas its latest edition (2022) switched to SPM acronym and defined strategic portfolio management technology as helping strategic portfolio leaders ensure enterprise-wide strategy-to-execution alignment and adaptation. Because projects move businesses and products forward, increased attention to request and service desk management has become more important to the end-to-end PPM model.

To reflect the importance of communication, the third edition of the *Standard for Portfolio Management* (PMI, 2013a) added a new knowledge area on Communication Management which outlines two processes, namely:

- Develops the portfolio communication management plan
- Manage portfolio information

The fourth and last edition (2017) replaces the Communication Management Knowledge Area as a Process Group by a Shareholder Management one (p 100 and p 103) with the following processes:

- Overview
- Guiding Principles
- Definition and Identification of Portfolio Stakeholders
- Analysis of Portfolio Stakeholders
- Stakeholders Engagement Planning
- Identifying Communications Management Approaches
- Manage Portfolio Communications

IMPLEMENTING A STRATEGIC PORTFOLIO MANAGEMENT COMMUNICATION

Recognize Maturity Levels

The Ancient Greek aphorism "know thyself" indicates that each organization must become conscious of where it stands in terms of communication around portfolio management in order to be in a position to move to the next maturity level, if deemed necessary and affordable. The level of maturity of portfolio management in the organization will influence the way communication is organized. Bayney & Chakravarti (2012) outlines five maturity levels in the People competency domain suggesting levels of maturity pertaining to communication and presentation (p. 156):

- Level 1 or the Ad-hoc level is characterized by using ineffective communication that may result in duplication of efforts across divisions. Decisions are not communicated effectively.

- Level 2 or the Basic level recognizes the need for a communication plan for portfolio management. The organization is working toward standardizing communication.
- Level 3 or the Standard level implies that a communication plan has been prepared and is used for portfolio management.
- Level 4 or the Advanced level has decisions communicated throughout the organization, with open communication being the norm.
- Level 5 is known as the Center of Excellence (CoE) level in which the organization's culture is collaborative and communicative. People are encouraged at all levels to submit ideas to foster continuous improvement.

Likewise, Bayney & Chakravarti outlined five maturity levels in the Strategy competency domain suggesting levels of maturity pertaining to collaboration (p. 158):

- Level 1 or the Ad-hoc level is characterized by projects initiated based on division needs and without regard to the impact on other divisions.
- Level 2 or the Basic level recognizes the need for identification of cross-functional synergies to achieve an enterprise focus.
- Level 3 or the Standard level implies that synergies, overlaps, and conflicts in each division's portfolio are discovered and resolved. Moreover, cross-functional opportunities are identified.
- Level 4 or the Advanced level has projects proactively reviewed to identify synergies, overlaps, and conflicts. Divisions collaborate to resolve them.
- Level 5 is known as the Center of Excellence (CoE) level in which portfolio planning, execution, and review are performed as a partnership.

Standardize and Simplify Communication Processes

Whatever methodologies and enablers are chosen to streamline portfolio management processes, they must be tailored to the organization and properly communicated to the various parties involved. In highly competitive markets, the need for trustworthy information becomes critical. The challenge for stakeholders is not obtaining more information but obtaining the right information. The system put in place must allow people to access, discuss, and share reliable and up-to-date information. An important step is to create and disseminate a simple formalized process for evaluating potential and current projects based on alignment to corporate strategic and financial goals. However, it is crucial to establish a process that requires every potential project to meet pre-determined criteria for acceptance giving executives confidence that they are making data-driven decisions on the greatest business value brought in and are not just appeasing influential stakeholders.

There are common questions that should be discussed upfront:

- What are the high-level objectives of the project? It's not uncommon for a project to morph into something very different from what was originally intended. Specifically identifying the desired outcome of every project helps project teams, sponsors, and stakeholders remain focused.
- What are the estimated costs of the project; what are the anticipated rewards? Without the answers to these questions, it is difficult to determine if the proposed project will provide any business value, let alone the greatest business value.
- Does the proposed project align with the mission, vision, and values of the organization? Individual projects must represent the execution of strategic direction and financial goals if the desired result is to maximize the return on investment (ROI).
- What risks are associated with pursuing the project under consideration? If risks can be identified and evaluated while the project is still in the consideration process, actions can be taken to mitigate risk and increase a project's probability of success.

In order to increase productivity and reduce costs, communication and collaboration around portfolio management will be eased by:

- Embedding rules and procedures in the process
- Developing and configuring user-friendly, natural, and enjoyable artifacts
- Creating repeatable work processes using enterprise templates
- Providing users with a means to update status with minimal impact on their work day

- Giving business leaders immediate access to business intelligence they need to make strategic decisions
- Gathering lessons learned geared toward continuous improvements
- Using cloud-based solutions whether these are public, private, or hybrid

Connect and Engage with People

The communication system deployed shall reduce the information overload by cutting out excess tools and connecting collaboration with work whether partners are or are not within corporate fences; in a nutshell, more context, less meetings. Connecting and engaging with people implies removing or at least reducing barriers that can retard or distort effective communication. Common barriers are technical and security layers, filtering, selective perception, emotions, language, silence, communication apprehension, etc.

- If the level of user involvement depends of the type of project or activities (Wysocki, 2009), being the ultimate communication piece between customers and implementers, the program and project managers shall be trained and possibly certified for using similar terminology, most often a blended version of a globally recognized methodology. They shall be invited to actively engage in topics of interest in dedicated forums.
- In order to avoid a mismatch between the classical paper-based, non-disclosure agreement (NDA) suitable to asynchronous communication channels, such as e-mail and the access to modern synchronous environments such as the professional social networks, it is crucial to have a process in place to safeguard the organization for improper use of information. Ideally, NDA signatures shall be embedded in a two-step authentication process for eluding paperwork.
- Most enterprise work scenarios require documents of various shapes and sizes at many different stages of work, for collaboration, approvals, and sharing. Documents shall be an integral part of the business and portfolio management life cycle. Teams should not waste time and resources inefficiently managing the creation of documents and digital assets from beginning to end. Documents shall be organized and accessible from a central location by managers, team members,

external stakeholders, or any third party contributors keeping all discussions, questions, comments, and versions in the context of the document throughout its life cycle.

Increase Visibility and Collaboration

An effective portfolio management communication system shall provide all levels of the organization with visibility and insight into the truth about workloads, dependencies, and when things will really be done. Independently of the channel, information is conveyed through managers, and senior managers shall get real-time visibility into who is working on what and how to justify the resources used, and the ones still needed. In order to make informed decisions, executives, managers, and project teams with the data shall be able to securely access standard dashboards as well as configurable reports whether they are on company premises or traveling for business.

To maximize impact and relevance as a CoE, the EPM office shall dispose of a properly designed communication room composed of wide screens displaying live portfolio information 24/7; a specialized library gathering references pertaining to portfolio, program, and project management; as well as video facilities to record and disseminate achievements and lessons learned.

SUMMARY

Once a senior manager working for a huge global corporation was asked what the most important topic was to address in portfolio management. His interviewer was indeed expecting him to answer the adherence to processes. He opted, however, for communication. According to that leader, despite the plethora of methodologies, frameworks, platforms, tools, and techniques that can be used or not depending on the context, especially in an organization so big, diversified, and globally distributed; communication is unique; as such, it is not an option.

Somehow communication transcends its environment by allowing collaborators to address strategy, performance, governance, and risks topics aligning operations and portfolios of programs and projects to achieve overarching business objectives.

MIND MAP

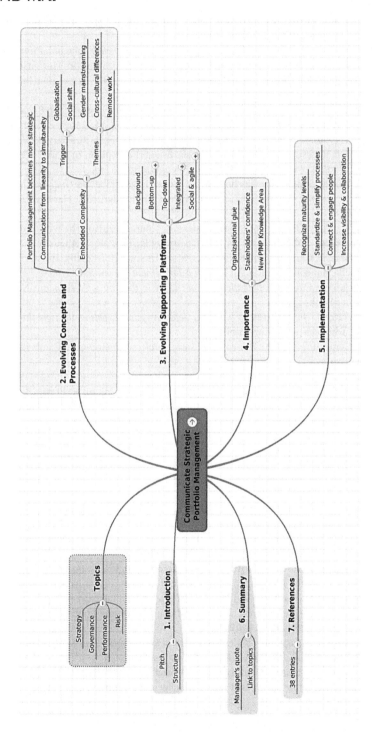

REFERENCES

Aberdeen Group Analyst Insight. (2013). *Harnessing the power of next-generation communications*. Boston, MA: Aberdeen Group.

Aubrée-Dauchez, A. (2005). *Portfolio management toolset overview. Memorandum.*

Barnlund, D. C. (2008). A transactional model of communication. In Mortensen, C.D., *Communication theory*. 2nd edition. New Brunswick, NJ: Transaction Publishers, 47–57.

Bayney, R. & Chakravarti, R. (2012). *Enterprise project portfolio management: Building competencies for R&D and IT investment success*. Boca Raton, FL: J. Ross Publishing.

Berlo, D. K. (1960). *The process of communication*. New York: Holt, Rinehart, & Winston.

Brafman, O. & Beckstrom, R. A. (2008). The starfish and the spider. In search of the sweet spot. *Portfolio trade*. Chapter 8. London: Penguin Books Ltd.

Burston, P. (2012). Google's diversity head: Mark Palmer-Edgecumbe. *The Guardian*, March 15.

Chester, B. I. (1938). *The functions of the executive*. Cambridge, MA: Harvard University Press.

Corporate Executive Board Corporate Leadership Council (2010). The role of employee engagement in the return to growth. *Bloomberg Business Week*, August 13.

Cross, R., Gray, P., Cunningham, S., Showers, M. & Thomas, R. J. (2010). The startling cost of inefficient collaboration. *MIT Sloan Management Review 52*, 83–90.

Davis, S. & Meyer, C. (1999). *Blur the speed of change in the connected economy*. New York: Warner Books Edition.

Drucker, P. (2006). *The effective executive: The definitive guide to getting the right things done*. New York: Harper Collins Publishers.

Forrester Consulting (2010). *The case for social project management*. Cambridge, MA: Forrester.

Gallup (2017). *Stage of the American workplace*.

Krebs, J. (2009). *Agile portfolio management*. Richmond, WA: Microsoft Press.

LEADing Practice (2014). Value reference framework. *LEADing Practice*. Retrieved from: http://www.leadingpractice.com

Levin, G., Artl, M. & Ward, J. L. (2010). *Portfolio framework: a maturity model*. Arlington, VA: ESI, International.

Locker, K. O. (2001). *Business and administrative communication*. Burr Ridge, IL: McGraw-Hill.

Office of Government Commerce. (2011). *Management of portfolios*. Norwich: The Stationery Office.

Project Management Institute. (2006). *The standard for portfolio management*. Newtown Square, PA. Project Management Institute.

Project Management Institute. (2008). *The standard for portfolio management*. 2nd edition. Newtown Square, PA: Project Management Institute.

Project Management Institute. (2013a). *The standard for portfolio management*. 3rd edition. Newtown Square, PA: Project Management Institute.

Project Management Institute. (2013b). *PMI's Pulse of the profession in-depth report. The high cost of low performance. The essential role of communications*. Newtown Square, PA: Project Management Institute.

Project Management Institute. (2017). *The standard for portfolio management*. 14th edition. Newtown Square, PA: Project Management Institute.

Rich. (2011). Project management software – A brief history (Part 1). *EPM Live*. Retrieved from: http://epmlive.com/?s=Project+Management+Software

Robbins, S. P. & Judge, T. A. (2013). *Organizational behavior.* 15th edition. Upper Saddle Ridge, NJ: Pearson Education.

Roberts, K. H. (1984). Organizational communication. In Kast, F. and Rosenzweig, J. (eds.). *Modules in management.* Chicago: SRA Associates, p. 4.

Royer, I. (2003) Why bad projects are so hard to kill. *Harvard Business Review 81,* 48–56.

Shannon, C. E. & Weaver, W. (1949). *The mathematical theory of communication.* Urbana, IL: University of Illinois Press.

Stang, D. & Handler, R. A. (2013). *Magic quadrant for cloud-based IT project and portfolio management services.* Stamford, CT: Gartner.

Stang, D. & Henderson, A. (2022). *Magic quadrant for strategic portfolio management.* Stamford, CT: Gartner.

Weick, K. E. (1987). *Theorizing about organizational communication. Handbook of organizational communication.* Newbury Park, CA: Sage.

Wroblewski, L. (2011). *Mobile first. A book apart.* Paris: Groupe Eyrolles.

Wysocki, R. K. (2009). *Effective project management: traditional, agile, extreme. Client involvement vs the complexity/uncertainty domain.* 5th edition. Hoboken, NJ: John Wiley & Sons, Inc.

9

Project Portfolio Management and Communication

Wanda Curlee

Project Portfolio Management (PPM) is an area of project management that has become more formalized. Managing a group of projects was done before PMI instituted a portfolio standard. The portfolio practice had several names. The Project Management Institute (PMI®) acknowledged PPM in 2006 by publishing its first portfolio management standard. The standard is now in its fourth edition, published in 2017 (PMI, 2017).

This chapter will focus on the importance of communication by those responsible for the portfolio. By extension, the governance team and the stakeholders are also responsible for communicating with those involved in portfolio management. The governance team and stakeholders must communicate up and down the chain of command. The portfolio manager and team are also responsible for sharing with all parties involved and ensuring that project managers provide adequate project status.

There are several standards on portfolio management by several project management associations. They may have some sections that are in more detail and provide tools and templates and step-by-step instructions. Most do have a section on communication. The Project Management Institute's standard will be reviewed as a sample.

The PMI portfolio standard has a section on communication titled "Identifying Communications Management Approaches" (PMI, 2017, p. 70). Governance, infrastructure, management plan, reports, assets, and components are all areas in the section. The standard also has a section on "Manage Portfolio Communications" (PMI, 2017, p. 73). The part is relatively short and discusses stakeholder communication.

This chapter will discuss the various communication paths needed in PPM. Stakeholders are treated differently depending on where they fall

DOI: 10.1201/9781003315902-9

within the portfolio. Having many types of stakeholders increases the complexity of communications. This complexity will be reviewed in more detail. Portfolio managers need to ensure that communications are all relevant to strategy and the differences in communicating to senior executives, the portfolio team and project managers.

GENERAL COMMUNICATION

Communication with the advent of Covid became more difficult. Most companies' leadership found themselves having to have employees work from home. Working from home was a drastic culture change (Global Project Portfolio Management Market – Growth, Trends, COVID-19 Impact, and Forecasts (2022–2027), 2022). Company employees not used to working from home faced a different communication dynamic. As Covid has waxed and waned, companies have returned to the office, are hybrid, or stayed virtual. Another shift for employees and leadership on communication needs.

For portfolio managers and teams accustomed to a traditional office environment, switching to a hybrid or virtual environment may be difficult. There may be a lack of face-to-face or only video communication, which does not provide the advantage of body language. Body language accounts for 80% of communication (Pal Singh & Singh, 2005).

The number of communication lines further complicates visual clues to communication. The portfolio manager and the team, along with the project managers, need to remember the formula for the lines of communication, $(N*(N-1))/2$. The number of people to communicate with is N (PMI, 2021).

Communication is fundamental to decision-making (Enoch, 2019). The portfolio manager must be able to synthesize the information from the team and project managers. Then the data is analyzed as to how it affects the portfolio. Recommended portfolio updates are communicated to the portfolio governance team for it to make decisions. These decisions are again analyzed and communicated to the concerned project managers. Decision-making relies on communicating effectively up the chain of command and down to those doing the work.

Is trust earned or given? Do you start a person at 100% or do they start at zero and have to work their way up? Trust is easy to lose and then tricky to regain (Kunnel, 2021). Is trust different in social media and video conferences? Scholars have difficulty understanding what a person feels when there is trust and when there is none (Kunnel, 2021). The section on trust will discuss how it is crucial for the success of a portfolio and ensure the strategy is met.

Strategy is the reason for establishing a portfolio. The portfolio should oversee the projects, programs and operations and focus them on the company's strategic goals. There needs to be a balance of cost, risk, and realized benefits. The portfolio team needs to understand the various stakeholders to manage the portfolio.

Technology will assist the portfolio team in communicating with stakeholders. Communicating with technology is essential today because of the Covid pandemic. Many now work in remote environments, and this type of communication may be foreign to many. The portfolio manager and the team need to understand the technology and when it is appropriate to use. Often, it may just be right to call, especially for issues with a quick answer.

Decision-making, trust, strategy, and technology will be reviewed in more detail.

DECISION-MAKING

The portfolio manager constantly is reviewing the projects, programs, and operations. How to decide which ones come into the portfolio. Do the components support the strategic goals of the company? Are the portfolio balance cost, risk, and benefits realized? Does the portfolio's risk tolerance meet the company's risk tolerance? Should the components be accelerated, delayed or killed?

Enoch (2019) aptly noted that a portfolio manager that does not make the correct decisions for the portfolio's components is directly related to communication. He further stated that three items need to be the focus. They were as follows:

1. Understand how the portfolio meets the company's goals;
2. Develop a process to evaluate the possible components to include and how each relates to the company's strategic plan; and
3. Develop lessons learned to enhance the portfolio.

(Enoch, 2019)

Wu and Chatzipanos (2018) stated that proficient portfolio managers are adept at soft skills. One of the soft skills includes decision-making (Wu & Chatzipanos, 2018; Enoch, 2019; Vacík, Špaček, Fotr, & Kracík, 2018). Vacík, Špaček, Fotr, and Kracík (2018) note that the portfolio needs to be driven by the company's leadership. The PfM needs to understand what leadership wants from the portfolio. This understanding requires in-depth communication.

Vacík, Špaček, Fotr, and Kracík (2018) understood that a portfolio manager needed to know that the portfolio not only needs to be balanced but also driven by the leadership's wants. The wants can be the multicriteria of the project or the net present value of the portfolio (Vacík, Špaček, Fotr, and Kracík 2018). Annosi, Marchegiani, and Vicentini (2020) emphasized that decision-making must happen across the organization. The portfolio team and the portfolio manager must communicate with project managers, business unit presidents, and other organizational leaders.

The portfolio team will have to take the information understood from the organization's various constituents and communicate it with each other. The communication by each team member has to be concise and precise. The others, including the portfolio manager, need to have the ability to paraphrase the message back to the team. The paraphrasing helps ensure that the group hears the same information, even though each team member was not present at the interviews.

Margolis (2004) noted that patients retain about 25% of what they hear from their doctors. Extrapolating this percentage to a business environment is plausible as those in the team may not have an in-depth knowledge of the business topic needed for the portfolio. It is of utmost importance that once the portfolio team exchanges information from the interviews, they listen to questions from the portfolio manager and others in the team. Listening will help each member grasp if they are communicating effectively. The team will have to decide if they need to follow up with the interviewee to clarify information.

Once the team analyzes the information gathered, the portfolio is created or updated. The team needs to relay to the portfolio manager why certain decisions were made and why the portfolio has the projects, programs, and operations within the portfolio. When the portfolio is updated, the team must explain why projects were left as is, accelerated, slowed down, or deleted. The portfolio manager needs to present the information to the portfolio governance team. The portfolio governance team may ask difficult questions to which the portfolio manager should have answers or at least an explanation (Lappi, & Kujala, 2019). The portfolio manager may need to take questions back to the team, but this needs to be the exception rather than the rule.

Finally, the communication needs to work in reverse. The portfolio manager communicates the team governance team's decisions. The team updates the portfolio as needed. Updates are then shared with project managers, business unit presidents and other leaders. The process is continual; as one cycle ends, the next one begins.

None of the decision-making and communication can be done without trust. Wrong decisions will lead to a lack of trust and a hesitancy to communicate up and down the chain of command (Wu & Chatzipanos, 2018).

TRUST

Trust in a relationship goes both ways. Trust is easier to maintain than it is to recover once it has been broken. PPM is a relationship on many levels. Some of these levels are as follows:

1. Between portfolio team members;
2. Between team members and the portfolio manager;
3. Between team members and project managers;
4. Between portfolio manager and project managers;
5. Between the portfolio governance team and the portfolio manager;
6. Between the portfolio manager and company leaders;
7. Between the team members and organizational leaders; and
8. Many others.

No matter where the relationship lies, maintaining trust will enhance the credibility of the portfolio and will keep communication open between all parties.

Within the project management profession, no one will deny that the portfolio communications are simple. As more companies formally adopt PPM, these organizations will realize the importance of trust. Each portfolio manager may use one or various methods to increase trust. Some of the types of leadership and communication that may be used are as follows:

1. Walking around and speaking with members of the team (only in traditional brick and mortar settings);

2. Contacting team members by having a video-teleconference every week (nominally in remote settings);
3. Servant leadership, where the portfolio manager is there to help the team member succeed;
4. Complete transparency with all stakeholders; and
5. Using technology to make the status of the portfolio transparent to all in the organization, even driving down to individual projects.

There are other ways to enhance trust; each portfolio manager needs to understand what works for them.

There are three models of communication. These are as follows:

1. The three views of communication;
2. Berlo's SMCR linear model of communication; and
3. The Perceptual Filter.

The oldest model is the three views of communication. This model views communication as an interaction, communication as action, and communication as a transaction (Fox, 1999). The other two models' foundations are based on the three views of communication.

Berlo's model is referred to as SMCR. The "S" is the sender; the "M" is the message the sender wants to communicate with the other party; the "C" is the channel. It is how the sender encodes the message (think language, email, virtual reality); and the "R" is the receiver, or the person who has to decode the message (Fu & Lv, 2020). Berlo's model assumes that communication is always the same. It does not take into account noise during transmission. Noise may be a distraction, no fluency in the language being used, lack of visual clues, chatting on social media, or biases, among other interferences.

The last model is the Perceptual Filter. This communication model considers many issues that disturb the sending and receiving of a message (Laing & Bergelson, 2020). Hoffner (2015) aptly points out that each generation has its form of communicating. Those various perceptual filters include the following (Hoffner, 2015):

1. Factual – This person focuses on the message being conveyed;
2. Self-revealing – This person is listening for the hidden message;

3. Affiliation – This person is looking for the "accusation and allegations" (Hoffner, 2015); and
4. Appeal – This person is listening for the message that can make me look better.

With these barriers, it is understandable that the portfolio leader and those on the portfolio team must understand how stakeholders listen to the message being conveyed. The portfolio leader needs to understand those that are direct reports and how they listen. The portfolio manager should be concise in the message, transparent, and watch for work-life balance and any accusations. The leader must ensure that they take responsibility for any failures and that the team or team member receives the credit for successes. This leadership and listening will increase trust with various perceptions of the portfolio team.

The portfolio team needs to trust one another and the portfolio manager. As noted in the above paragraph, the portfolio manager who supports the team will establish trust. When the team supports each other and becomes true to their word, this will help to develop and maintain trust.

Stakeholders outside the portfolio team include leadership, the portfolio governance group, and project managers. The portfolio team will interact with each of these stakeholders. When listening, the "Perceptual Filter" and the generation needs to be understood (Lappi, & Kujala, 2019). The younger generations may prefer to discuss via text. In contrast, older generations may want to discuss face-to-face or via voice. The remote nature of projects and the increase in technology have enhanced both communication styles. The section on technology will address technologies that assist with communication.

Communicating via technology still has its limitations with understanding body language. Body language accounts for 80% of communication understanding (Pal Singh & Singh, 2005). Some believe video technology takes care of this problem. Video technology relies on bandwidth, clarity, lag time, and voice coordinated with body movement, among other items. There is the issue of keeping a person's attention during video conferences. Due to social media, attention spans have become shorter. Instant gratification is wanted. The leader of the video conference needs to publish an agenda, keep the meeting on track, and keep the attention of all. All decisions need to be documented and confirmed with the parties affected.

Chatting via text, especially when a decision is made, as with video conference meetings, the decision must be documented and shared with those in the chat. The purpose is to minimize misunderstandings of the text message.

Portfolio team members and the portfolio manager need to understand the organization's culture and the method used to communicate with stakeholders. The portfolio manager and the team may have to alter their communication style while understanding the receivers' perceptual filters.

STRATEGY

To deliver the portfolio, a successful set of projects, programs, and operations is imperative (PMI, 2017). The portfolio team initially selects the projects considering cost, risk, benefits realization, and resources. Resources and costs typically drive the composition of the portfolio (Wu & Chatzipanos, 2018; Vacík, Špaček, Fotr, & Kracík, 2018). The initial portfolio and updates to the portfolio are communicated to the portfolio governance for final approval.

Strategy, by its nature, may be challenging to explain and select projects that drive the company's strategic goal(s). Hadjinicolaou, Kader, and Ibrahim (2022) state that the portfolio and the team should be flexible to meet company goals. Industries change rapidly, and while strategy does not alter often, it may vary slightly or radically when the industry changes.

Communication is essential with strategy. The governance team must ensure that it articulates how to measure the portfolio (Joseph, 2019). The portfolio manager should ask precise questions and attempt to resolve differing opinions. When the governance team has differing opinions, the portfolio manager should press the head of the governance team to make the final decision (Vacík, Špaček, Fotr, & Kracík, 2018). The portfolio manager should communicate findings to achieve the strategic goals upon a final decision.

The portfolio team then assembles the portfolio, if new, or updates an existing portfolio (Chatterjee, Hossain, & Kar, 2018). The portfolio manager must understand how these projects meet the strategic need while others do not. The portfolio manager must communicate the new portfolio or updates proficiently to the governance team and the affected project managers.

The portfolio manager needs to demonstrate transparency to the entire company. There are various communication methods the portfolio manager may use (Vacík, Špaček, Fotr, & Kracík, 2018). Technology may allow the organization to view the portfolio and drill down to the projects and how the projects are performing in real-time. Another method is to have conferences for any in the company to attend and ask questions. Newsletters are another communication method. The article within the newsletter should pique readers' interest while soliciting comments. The portfolio manager must respond to the queries to maintain interest in the portfolio (Annosi, Marchegiani, & Vicentini, 2020). As the portfolio's interest wanes, there may be an issue with the portfolio. Understanding the lack of interest must be addressed quickly.

Project managers need to understand the company's strategic goals. They need to make sure the project continues to meet the objectives. Projects can change over time. When this happens, the project manager must communicate with the portfolio team or the manager.

Strategy is the center of portfolio management. The portfolio manager must constantly review the best way to achieve the strategy while staying flexible enough to move with the industry and the company goals. The portfolio team and project managers need to understand the strategic goals to focus the portfolio. Finally, the company's leaders and portfolio governance team need to be transparent and upfront with the portfolio manager.

TECHNOLOGY

Technology, seldom, if ever, is the answer to communication problems. Technology may help with communication issues, but much effort is needed to incorporate the technology into processes and procedures.

Portfolio management software has matured but still does not meet all needs. Portfolio managers must consistently remind all stakeholders of the various messages they may receive. As the technology sends reminders to do things, some will view the messages as a nuisance and begin to ignore them.

More advanced project portfolio software will send reminders to those ignoring messages and the portfolio manager or team member. That is when a personal touch is needed. The portfolio manager needs to approach nonresponding individuals and understand why. It may be as simple that there was a misunderstanding of the messages. The person may be new and did not realize the importance. Others may know the information provided elsewhere and why they must respond twice. The portfolio manager needs to address these issues by instituting training and speaking with the information technology department.

Artificial intelligence (AI) may be a significant source of information for the portfolio. Disruptive technologies, such as artificial intelligence, can provide data from throughout the company that must be instituted in the portfolio and project software. This would be the responsibility of the portfolio manager to coordinate with the information technology department. The portfolio manager may have to create a business plan as it will possibly take resources away from the portfolio. Things to consider are as follows:

1. How will AI drive to better the company's return on investment?
2. How will having more information on projects help the portfolio have better results?
3. Do the benefits outweigh the negatives?

Technology can help the portfolio but never be led by a software salesperson who states that the software will resolve all communication issues. It will not.

CONCLUSION

Portfolio management continues to evolve the practice. It is up to each portfolio manager to stay current with changes, whether technology, agile, more accurately selecting projects to meet strategy, or many other developments. No matter what area is updated, communication will be central to all areas of portfolio management.

This article reviewed communication in various portfolio areas, including decision-making, trust, strategy, and technology. Decision-making is central to the reason a portfolio is created. Without proper communication, the portfolio could face less than optimal results. The business, in the long run, would suffer.

Trust is important to all relationships. A portfolio is a relationship between the portfolio team and the stakeholders. Without trust, the team will be unable to discern the critical information from that which does not provide value. The team has to provide a value-add to leadership. If the team cannot communicate well, minimal value will be added.

Strategy is the reason for the portfolio. The portfolio team and project managers must understand the company's strategic goals and how its leadership wants to achieve them. Are projects selected to increase the net present value of the portfolio, or are projects selected by project criteria? Depending on the answer, it will drive the projects within the portfolio.

Technology is never the complete answer to communication. Technology can be an assistance with finding the data needed to communicate with the organization, no matter at what level the person is. Face-to-face communication may be required even when technology is wholly integrated within the company.

People desire communication. Depending on the generation of the employee, the communication may be different. Some may be better with text, others with email, and others with a phone call. The portfolio team needs to understand what each stakeholder prefers to communicate effectively.

REFERENCES

Annosi, M., Marchegiani, L., & Vicentini, F. (2020). Knowledge translation in project portfolio decision-making: the role of organizational alignment and information support system in selecting innovative ideas. *Management Decision*, 58(9), 1929–1951. https://doi.org/10.1108/MD-11-2019-1532

Chatterjee, K., Hossain, S., & Kar, S. (2018). Prioritization of project proposals in portfolio management using fuzzy AHP. *Opsearch*, 55(2), 478–501. https://doi.org/10.1007/s12597-018-0331-3

Enoch. (2019). Project portfolio management, second edition: A model for improved decision making. In *Project portfolio management*, 2nd Edition. Business Expert Press.

Fox, D. (1999). Views and Models of Communication. http://communication-theory.freeservers.com/custom.html

Fu, Q. & Lv, J. (2020). Optimal design of information elements in virtual reality system based on TOPSIS. *Journal of Physics. Conference Series*, 1654(1), 12073. https://doi.org/10.1088/1742-6596/1654/1/012073

Global Project Portfolio Management Market – Growth, Trends, COVID-19 Impact, and Forecasts (2022–2027): The Global Project Portfolio Management Market (henceforth referred to as the market studied) was valued at USD 4,908. 4 Million in 2021, and it is expected to reach USD 6,284. 3 Million by 2027, registering a CAGR of 4. (2022). NASDAQ OMX's News Release Distribution Channel.

Hadjinicolaou, N., Kader, M., & Ibrahim, A. (2022). Strategic innovation, foresight and the deployment of project portfolio management under mid-range planning conditions in medium-sized firms. *Sustainability*, 14(1), 80. https://doi.org/10.3390/su14010080

Hoffner, L. (2015). Staff Communication: Control, Filters, and Perceptions. *Camping Magazine* (Online), URL: Staff Communication: Control, Filters, and Perceptions | American Camp Association (acacamps.org),

Joseph, M. (2019). Effect of Strategic Alignment and Portfolio Governance on IT Effectiveness: A Correlational Study. ProQuest Dissertations Publishing.

Kunnel. A. (2021). *Trust and communication*. Peter Lang International Academic Publishing Group. https://doi.org/10.3726/b17879

Laing, C. & Bergelson, E. (2020). From babble to words: Infants' early productions match words and objects in their environment. *Cognitive Psychology*, 122, 101308–101308. https://doi.org/10.1016/j.cogpsych.2020.10130

Lappi, A., & Kujala, J. (2019). Project governance and portfolio management in government digitalization. *Transforming Government*, 13(2), 159–196. https://doi.org/10.1108/TG-11-2018-0068

Margolis, R. (2004). What do your patients remember?. *The Hearing Journal* 57(6), 10, 12, 16–17. https://doi.org/10.1097/01.HJ.0000292451.91879.a8

Pal Singh, V. & Singh, B. B. (2005). Communication skills for effective transaction of curriculum in futuristic classrooms. *i-Manager's Journal on School Educational Technology*, 1(2), 25–29. https://doi.org/10.26634/jsch.1.2.925

Project Management Institute. (2017). *Standard for portfolio management*. 4th Edition. Newtown Square, PA: Project Management Institute.

Project Management Institute. (2021). *A guide to the project management body of knowledge* (PMBOK® Guide). 7th Edition. Newtown Square, PA.: Project Management Institute.

Vacík, E., Špaček, M., Fotr, J., & Kracík, L. (2018). Project portfolio optimization as a part of strategy implementation process in small and medium-sized enterprises: A methodology of the selection of projects with the aim to balance strategy, risk and performance. *E+M Ekonomie a Management*, 21(3), 107–123. https://doi.org/10.15240/tul/001/2018-3-007

Wu, T. & Chatzipanos, P. (2018). *Implementing project portfolio management: a companion guide to the standard for portfolio management*. Newtown Square, PA.: Project Management Institute, Inc.

10

Delivering Organizational Value in the Zone of Uncertainty

Lynda Bourne

The dynamics of the global economy and the increased complexity of delivering value to stakeholders have had consequences for organizations. Uncertainty about global events such as the COVID pandemic, effects of climate change and constant wars in different parts of the globe influence organizations' strategy and views about what success means. Before the advent of these global events and experience of their impacts, there was a view that managing an organization's work could be straightforward – if only we could get the processes and controls right! The formidable mixture of multiple developments and operational activities at differing stages of delivery was seen as a set of obstacles that could be overcome with more and better planning and controls. Processes and controls are only part of the story; without an understanding of the nature and sources of the complexity we all encounter, the efforts of everyone involved will be squandered.

The picture post-2020 is quite different! The consequences for individuals, communities and nations of COVID have added further uncertainty and complexity to an already complex situation. This complexity will affect an organization's ability to develop, implement, and control their portfolios, programmes and projects. The source of this complexity is a combination of technology, organization processes introduced to 'balance' the organization's work and relationships between stakeholders. This chapter will focus on the third of these factors: management of the relationships between the organization and the many stakeholders who can affect or who are affected by this work – all in a climate of increasing uncertainty.

The more the organization 'balances' the portfolio through selection of work to provide a broader range of opportunities for organizational value,

DOI: 10.1201/9781003315902-10

the more complicated the mix of knowledge, action, and relationships becomes. These are all in the domain of the *stakeholder factor*. People apply knowledge (and unconscious bias), people act (and make decisions) and people must work with others to achieve this value (relate). It is people – *stakeholders* – who make the design decisions and purchase decisions; people are responsible for the approvals and prioritizations. It is people who may or may not comply with, or who misinterpret, the processes; and it is people who implement the solutions. At any step in the organization's prescribed procedures and processes, the intention of the strategy may be diverted (or subverted) through conflicting goals, uncertainty, lack of understanding of the importance of following a process or just plain apathy.

The principles and processes, plans and measures that are put in place in organizations to support effective management of portfolios, pro-grammes and projects are important not only for consistency of approach but also to help build a culture within the portfolios through common symbols such as shared meaning through language and team tasks (Hofstede, Hofstede, & Minkov, 2010). The effectiveness of the application of these fundamental building blocks is enhanced through a focus on stakeholder relationships.

The chapter is organized as follows: first a description of the *Zone of Uncertainty* – the optimistic view and the reality is given. The next section describes potential remedies for the disruption and disappointment that occurs in the Zone, principally related to improving relationships amongst stakeholders. Any discussions of portfolio are based on the definition that portfolio management is *'doing the right things'* – effectiveness; *'doing things right'* is execution enabled by project management/work manage-ment (efficiency) (Cooke-Davies et al., 2011).

UNCERTAINTY – TIME TO RETHINK?

The future is uncertain! We don't know what is going to happen next. *Will we make it to the next milestone? Will my contract be renewed? Will unex-pected storms ruin my outdoor wedding?* We can rely on the predictions of weather forecasts or cumulative scientific knowledge, but mostly we just hope for the best, only sometimes preparing for the worst. Previously, we chose to consider an absence of certainty as unremarkable and therefore no cause for concern *if it does not affect us or our plans.* Under normal circumstances we tend to be influenced by inherent biases such as Optimism Bias and are more inclined to treat risk management in many projects as a box-ticking exercise. Post-2020 we are more affected by global events: how millions of people died during the COVID pandemic; how economies tanked; and how we had to develop new ways of thinking and dealing with lives and futures. Is it also time to rethink our approach to risk identification and management?

What is Risk, Really?

Risk has been previously defined as *uncertainty that matters* or uncertain events that will have a positive or negative effect on completion or success of an activity if it occurs. In this chapter, the definition is modified: *risk is uncertainty that matters to you* (or your project, organization, or community). The pandemic and its economic aftermath affected the whole global population, and so, the disruption to our lives, the fear of infection was everybody's uncertainty. Different people or groups reacted to this uncertainty in different ways, depending on their risk appetite, their access to accurate information and also their personal views of the role of governments.

Risks are often categorized as 'known unknowns' (not being able to find an expert for the team at the right time) or 'unknown unknowns' (a pandemic). Pre-pandemic, project risk identification and management were often seen as an early project ritual, a box to be ticked, the resulting documentation then filed away with other project archive documents. In other, better managed projects, risk management had been seen as important as an act of preparedness and diligence. However, communication of the consequence of risks or issues often needed to be modified to make allowances for management's alarm at the delivery of the 'bad news' or concern about the effect of risk treatments on the project budget.

The anxiety that results from risk realized through unusual examples of uncertainty can be viewed as a continuum, tracing the awareness of the consequences of the issue (no longer a risk) and the consequent anxiety over time, as:

- News of other disasters replaces the original horror and immediacy of an event. For example, in 2022 the disaster of flooding in Australia's eastern regions and the invasion of Ukraine relegated information about daily COVID hospitalizations and death to later parts of the news – if at all.

- More information becomes available – as scientists learned more about the new issues that beset us, they were able to explain what was really happening and what was being done to improve the situation.
- News about the progress of any global or local catastrophe was no longer 'new'. We had adapted to the changing situations and the consequences of actions being taken to keep us safe.

The *Zone of Uncertainty*

The *Zone of Uncertainty* is the murky area between an organization's view of how to achieve business success and the actual work that delivers value to the organization; this includes project work. In this Zone the hopes and dreams of the organization's strategists collapse under the weight of the complications and complexity of creating value, characterized by a perception of lack of control, lack of certainty and lack of direction and, right now, disruption and anxiety.

The Optimistic Approach

As part of the regular business planning cycle an organization defines its business strategy for a particular interval. From that strategy a clear set of strategic objectives should emerge; these strategic objectives can be converted into programmes and projects. Optimistic expectations anticipate that there is a straightforward path from milestone A to milestone B; all that is needed is the perfect combination of the right approvals, structures and processes in place (command) and then measures and reporting against those measures (controls). The transition from the business strategy to the output of the project occurs in the Zone.

The Zone is the highly complex and dynamic region between an organization's strategic vision and the portfolios, programmes and projects created to deliver that vision. It includes all the initiatives introduced by senior (or middle) management to deliver value. Zone activities include any aspect of management to ensure that all the work:

- Delivers the organization's strategy in a balanced way;
- Is appropriately funded and resourced;
- Is tracked and controlled to ensure progress (time, cost and scope) is delivered according to the schedule and any necessary managed adjustments are implemented;

FIGURE 10.1
Idealized view of how projects will deliver the organization's business strategy.

- Delivers the benefits outlined in the business case – the project's outcomes.

Figure 10.1 describes the idealized Zone. Clear management direction and defined outcomes are shown as unclouded by culture or unconscious bias and without any divergence or conflict. Clear transmission of strategic (organizational) visions into tactical (project) objectives ensures that projects will be delivered to the required time, cost and quality. This view supports the assumption that there will be universal agreement from all stakeholders that the chosen path is the best and only path for delivery of the outcomes of the portfolio, programme, or project.

The Reality
Figure 10.2 shows the more realistic picture of what happens in the Zone. Management's expectations remain unwavering[1], but the outcomes are not so predictable. The simple direct line of assumption between the executive's articulation of strategic objectives and the project work that is approved and resourced will most likely deviate. The path for delivery is affected by unexpected events, and senior stakeholders and the project team will react to try to regain control of the delivery of the objectives. Often these adjustments cause more disruption within the project and its relationships with other projects.

FIGURE 10.2
A more realistic view of the transition from business strategy to project.

The Project Management Institute's (PMI) *Standard for Project Management* (2021, p. 7–12) describes the components for value delivery:

> … [P]ortfolios, programmes, projects, products and operations…can be used individually or collectively to deliver value. Working together these components create a system for delivering value that is aligned with the organization's strategy. … A system for value delivery is part of an organization's internal environment that is subject to policies, procedures, methodologies, frameworks, governance structures and so forth. That internal environment exists within a larger external environment that includes the economy, the competitive environment, legislative constraints. … Outcomes create benefits and … benefits create value….
>
> The governance system works alongside the value delivery system to enable smooth workflow, manage issues and support decision making.

Methodologies, processes and practices that define how best to manage portfolios, programmes and projects are well defined and provide guidelines on the 'what must be done'. The process of delivering project outputs or delivering value to the organization through portfolio management assumes adherence to these processes. However, success could not be assured without considering the *stakeholder factor*. Application of and compliance with consistent sets of processes can deliver value in the form of timely delivery of outcomes. Timely delivery leads to improvement of customer satisfaction and return business. It is still, however, important to read into the words *application* and *compliance* the notion that the practitioners need to be trained, and managers need to recognize the importance of investing in the implementation, training and monitoring the implementation of these processes. This role is best lodged within a PMO.

Complexity

There is no doubt that the work of portfolio, programme or project management is complicated[2] and often complex[3]. An organization's work will always be complicated; what will make it complex is the combination of technical complexity, the specific selection and management of work to balance portfolios and the web of relationships with the stakeholder community in the environment of unpredictability.

We live with unpredictability, often unconsciously, every day and every minute. We do not know what is going to happen next; we take it on faith that our plans for tomorrow or further into the future will be able to be executed. The best example of this faith is the optimism of the project, programme or portfolio *schedule*. The schedule is developed by the team, including experts, but is at best just a 'guesstimate' of the best way to deliver the outcomes within the time frame, budget and the available resources to achieve the greatest value for the stakeholders or the organization. Despite the best intentions of all concerned, and even with the experience of others who have done similar work before, the plan assumes that there will be only a few complications, and the contingencies and other risk responses planned in will be sufficient. But we know that unexpected events will occur, and the carefully developed plan will need to be re-thought time and time again. This is normal. What is not normal is when stakeholders believe the plan: believe that it is TRUTH. Managing the unrealistic expectations of stakeholders based on this belief is a key part of stakeholder relationship management. The area between the expectations of how the organization must plan, operate and control all the work and each project, programme and operational activity is a minefield of unpredictability and complexity; expectations that the planned actions and changes will occur in a linear fashion will inevitably lead to failure (in the eyes of someone).

The complexity of human relationships is seen in situations where:

- People with different interests, loyalties, cultures and interactions with one another are put together to deliver something:
  - Teams come together causing unpredictable behaviour particularly when there are differences in national, organizational or professional culture.
- Differences magnified by different perspectives and expectations of the people involved:
  - Those who implement, those who use the products, those who benefit from the outcomes of the project or those who establish a regulatory environment.
- Clients cause scope changes or delay important decisions.
- Managers react inappropriately to cost, schedule, scope or quality pressures.
- High-impact, low-likelihood risk events (issues) – 'unknown unknowns' and the reactions of individuals and groups to the uncertainty and unpredictability arising from such events.

Stakeholders

Anyone who has ever had to involve the stakeholder community of a project, programme or portfolio will be aware that the reactions of each individual can be unpredictable; the additional complexity of understanding which groups or individuals are important and supportive is complicated. This section will discuss the fundamentals of understanding and engaging stakeholders.

The concept of stakeholder existed long before management writers (in the West) took up the cause and adapted the word and concept for an organizational purpose. www.dictionary.com defines *stakeholder* in the following ways:

1. The holder of the stakes of a wager (*this was the original meaning of the word*).
2. A person or group that has an investment, share or interest in something, as a business or industry (this is now the generally accepted usage in the English-speaking business world).
3. In law: a person holding money or property to which two or more persons make rival claims.

The meaning and concept of *stakeholder* in the English-speaking world is hazy, with no common agreement on who are stakeholders. The various translations of *stakeholder* into other languages seem to indicate quite specific and differing ideas about the role and influence of stakeholders. The following list is derived from discussions with colleagues from different language backgrounds during the time that I was consulting with organizations in these countries[4]:

- In Spanish: *tenedor de apuestas (holding the wager)* and *partes interesadas (interested parties)*
- In French: *Des parties prenantes* (parties who are taking) and *intervenantes* (intervening)
- In German: *Beteiligten* (involved) and *Anspruchsgruppen* (who have a claim)
- In Dutch: *belanghebbenden* (having a stake)
- Paul Dinsmore (Dinsmore, 1999) referred to 'stakeholders' as the *'ones who have the beef'*. This seems to be a consistent view of 'stakeholder' in Brazil.[5]
- In Japan: 'stakeholders' are 'related people' or 'people sharing risk and profits'.[6]
- In China: 'stakeholders' are 'participants with related interest'.[7]

This brief and somewhat subjective analysis leads to the conclusion that there is no innate or generic meaning for stakeholders, and so any discussion about stakeholders in organizations needs to begin with a definition. This definition comes from PMI's *Standard for Project Management* (PMI, 2021, p. 31).

An individual, group, or organization who may affect, be affected by, or perceive itself to be affected by a decision, activity, or outcome of a project. Stakeholders also directly or indirectly influence a project, its performance or outcome in either a positive or negative way.

The strength of this definition of stakeholders is that it not only considers the ubiquity of stakeholders but also the diverse functions that they may have within the organization and outside it. It therefore assists in ensuring that no individual or group that fits the definition of stakeholder can be ignored. An additional strength in this definition is the focus on perceptions as a characteristic of stakeholder. Perceptions and expectations of stakeholders are important factors in successful stakeholder engagement and therefore to the organization's outcomes. Understanding what each key stakeholder expects to gain (or lose) from the outcomes of the work (either its success or failure) is essential to the success of the work and the perceptions of the stakeholder community. A stakeholder has a *stake* in the activity, portfolio, programme or project. It is important to consider the nature of a stakeholder's stake when defining a stakeholder's needs, requirements or how the individual or group can impact the organization's activities. This stake may be:

- An interest: a circumstance where a person or group is affected by a decision, action or outcome.
- Rights: legal – as enshrined in legislation; or moral – environmental, heritage or social issues.
- Ownership: legal title to real property, intellectual property or a worker's right to be compensated for his or her experience or application of specialist knowledge.
- Contribution in the form of knowledge or support.

A Methodology

It is not possible to measure the expectations or perceptions of people objectively and foolish to imagine that it is possible to guess or assume them. It is only possible to measure trends or changes in levels of stakeholder satisfaction or engagement (Bourne, 2014). A consistent approach – a methodology – for tracking any changes is critical. Such a methodology can record significant elements such as support, to provide a baseline of data about support or level of engagement. Changes in data about these elements subsequently gather will indicate improvements or otherwise in stakeholder attitudes; it will be a useful indicator of the success of the team's efforts at engagement.

The *Stakeholder Circle*® methodology is based on such a concept. Figure 10.3 shows the relationships between the activity[8] and its stakeholders. At the centre of the image is the activity. Surrounding the activity is the team, often overlooked in many stakeholder engagement processes. The community of stakeholders that has been identified as being important to the success of the activity *at the current time – time now* is pictured in the third circle. The outermost circle references potential stakeholders: those who may be important to the success of the work at a later stage. By differentiating current stakeholders and potential stakeholders, confusion

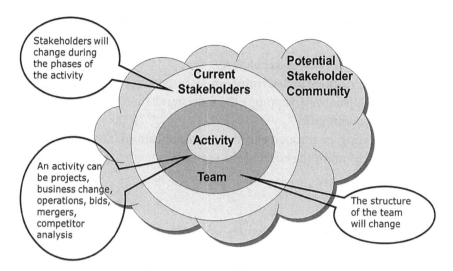

FIGURE 10.3
Stakeholder relationships.

about which stakeholders are important at any particular time and how best to manage the current relationships will be minimized; this approach ensures that planning for future relationships is managed effectively. The stakeholders in the outer circle must also be considered in risk management planning because they may cause the activity to be at risk of failure in the future. Alternatively, these stakeholders may need to be considered in an organization's marketing plans as potential customers.

Managing Stakeholder Relationships
The *Stakeholder Circle* is a five-step methodology based on the concept that any activity can only succeed with the informed consent of its stakeholder community (Bourne, 2012); managing the relationships between this community and the team will increase the chances of success. The team must develop knowledge about this community and based on that knowledge make judgements of the right level of engagement for each group. This information will help define the appropriate level and content of communication needed to influence stakeholder's perceptions, expectations and actions. It is important to emphasize that stakeholder relationship management is complex and cannot be reduced to a formula: each person is unique, and the relationships between people reflect that uniqueness and complexity and that attitude can change with changing situations.

The methodology consists of five *steps*:

- *Step 1*: Identify all stakeholders,
- *Step 2*: Prioritize to determine who is important,
- *Step 3*: Visualize (mapping) to understand the overall stakeholder community,
- *Step 4*: Engage through effective communications, and
- *Step 5*: Monitor the effect of the engagement.

Step 1: Identify

This step consists of three activities:

- Developing a list of stakeholders,
- Identifying *mutuality*:
  - How each stakeholder is important to the work, and
  - What each stakeholder expects from success (or failure) of the project or its outcomes (their expectations),
- Categorize: Document each stakeholder's *Influence category*:
  - *upward* (senior stakeholders), *downward* (the team), *outward* (public, government, suppliers, shareholders, etc.) and *sideward* (peers of the manager of the activity),

The output of this step will be a list of *all* stakeholders that fit the definition of stakeholder.

How Many Stakeholders?
Beware of STAKEHOLDER MYOPIA! Some organizational activities are large and complex and may affect many stakeholders. For example, construction of public facilities or national infrastructure projects will affect private citizens, landowners and the natural and historical environment. For such projects, there will be large numbers of stakeholders. There is often an unconscious limit within an organization on what a good number of stakeholders can be – this is stakeholder myopia. The team and management must understand that while the initial number of stakeholders identified may appear unwieldy or overwhelming, Step 2: Prioritize provides a structured and logical means to prioritize the key stakeholders for the current time.

Step 2: Prioritize

Most stakeholder management methodologies rely on an individual's (or the team's) subjective assessment of <u>who</u> is important. The approach adopted in the *Stakeholder Circle* methodology attempts to provide consistency in decision-making about stakeholders. It does this through a structured decision-making process where team members agree on and rate the characteristics of stakeholders to assess their importance to the success of the work relative to other stakeholders. *Step 2: Prioritize* provides a system for rating and therefore ranking stakeholders. The ratings are based on three aspects:

- *Power*: The power an individual or group may have to permanently change or stop the project or other work[9]
- *Proximity*: The degree of involvement that the individual or group has in the work of the team
- *Urgency*: The importance of the work or its outcomes, whether positive or negative, to certain stakeholders (their stake), and how prepared they are to act to achieve these outcomes (stake).

The team applies ratings to each stakeholder, for 1–4 for *power*, and *proximity* (where 4 is the highest rating) and 1–5 for each of the two parts of *urgency – value* and *action* (where 5 is the highest rating). The ratings support development of a ranked list of stakeholders who are important to the success of the work at a particular time of the life cycle.

Why Choose These Prioritization Attributes?

The three attributes of *power, proximity and urgency* are the essential elements for understanding which stakeholders are more important than others. The definition of *power* seeks to identify those who have power over the continuation of the work itself – or no power at all. *Proximity* provides a second way of identifying how a stakeholder may influence the work or its outcomes; regular, close and often face-to-face relationships will influence the outcomes of the work[10]. The immediacy of this relationship contributes to trust between members of the team, and more effective work relationships, as the team members develop better understanding of the strengths and weaknesses of those with whom they work on a regular basis. *Urgency* is based on the concept described in Mitchell, Agle, &

Wood (1997) whose theory described two conditions that may contribute to the notion of urgency:

1. Time sensitivity: Work that must be completed in a fixed time, such as a facility for the Olympic Games.
2. Criticality: An individual or group feels strongly enough about an issue to act, such as environmental or heritage protection activists.

In the *Stakeholder Circle, urgency* is rated through analysis of two sub-categories: the *value* that a stakeholder places on an outcome of the work, and the *action* that he or she is prepared to take as a consequence of this stake. The inclusion of *urgency* in the prioritization ratings balances the potential distortion of an organizational culture that identifies stakeholders with a high level of hierarchical power as most important. If *power* and *proximity* are the only measures, stakeholders such as the 'lone powerless voice with a mission', who can cause significant damage to successful outcomes if ignored, will not be acknowledged.

Step 3: Visualize – Mapping Complex Data

The objective of every stakeholder mapping process is to:

- Develop a useful list of *current_*stakeholders and assess some of their key characteristics.
- Present data to assist the team's planning for engaging these stakeholders.
- Reduce subjectivity.
- Make the assessment process transparent.
- Make the complex data collected about the stakeholders easier to understand.
- Provide a sound basis for analysis and discussion.

Presenting complex data effectively will be directly useful to two important stakeholder groups: the organization's management generally requires information in the form of lists, tables, pictures or graphics, whereas the project team responsible will need charts and graphics for analysis of the community to highlight potential issues. The mapping from the *Stakeholder Circle* fulfils all these requirements[11].

The Stakeholder Circle

Figure 10.4 is an example of how data gathered during *steps 1 and 2* of the *Stakeholder Circle* methodology is shown. In addition to this graphic the names of the stakeholders represented is printed alongside. Key elements of the *Stakeholder Circle* are as follows:

- Concentric circles indicate distance of stakeholders from the work of the activity.
- The size of the block represented by its relative length on the outer circumference indicates the scale and scope of influence of the stakeholder.
- The radial depth of the segment indicates the stakeholder's degree of power.

Colours[12] help the interpretation and indicate the stakeholder's *influence category* relative to the activity: Orange indicates an *upward* direction,

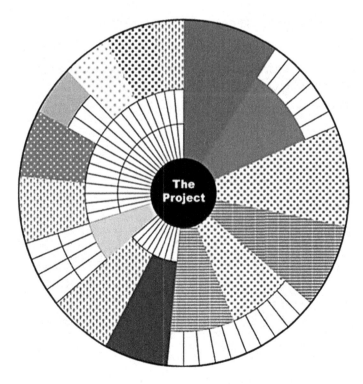

FIGURE 10.4
The *Stakeholder Circle.*

green indicates a *downward* direction, purple indicates a *sideward* direction, and blue indicates an *outward* direction.

Step 4: Engage

The fourth part of the *Stakeholder Circle* methodology is centred on defining engagement approaches tailored to the expectations and needs of these individuals or groups. The first step of this analysis defines the level of interest of the stakeholder(s) at five levels: from committed (5), through ambivalent (3), to antagonistic (1). The next step is analysis of the receptiveness of each stakeholder to messages about the project: on a scale of 5, where 5 is – direct personal contacts encouraged, through 3 – ambivalent, to 1 – completely uninterested. The third step is to identify the target attitude: the level of support and receptiveness to messages that would best meet the mutual needs of the project and the stakeholder. Figure 10.5 illustrates two stakeholders' engagement levels.

Step 5: Monitor Effectiveness of Communication

The matrix illustrated in Figure 10.5 becomes the engagement baseline and starting point for measuring communication effectiveness. A stakeholder's *attitude* toward an organization or any of its activities can be

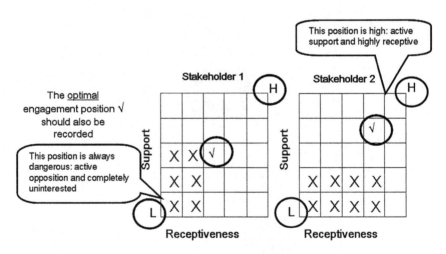

FIGURE 10.5
Stakeholder engagement profile.

driven by many factors, including whether involvement is voluntary or involuntary; whether involvement is beneficial personally or organizationally; or whether the level of a stakeholder's investment either financial or emotional in the activity. If the individual's or group's stake in the activity is perceived to be beneficial, or potentially beneficial to them, they are more likely to have a positive attitude to the activity and be prepared to contribute to the work to deliver it. If on the other hand, they see themselves as victims, they will be more likely to hold a negative attitude to that activity. Any assessment of *attitude* will need to consider the following elements:[13]

- Culture of the organization doing the work,
- Identification with the work and its outcomes or purpose,
- Perceived importance of the activity and its outcomes to the stakeholder; and
- Personal attributes, such as cultural background, personality, or position in the organization.

Any relationship requires constant work to maintain; this applies to family relationships, friendships, management of staff, and maintenance of professional networks. Relationships between an organization and its stakeholders are no different. The team must understand the expectations of all the important stakeholders and how can they be engaged through targeted communication to maintain supportive relationships and to mitigate the consequences of unsupportive stakeholders for the benefit of the organization and its activities.

The Communication Plan
The basis for an effective communication plan is defining for each stakeholder:

- The *purpose* of the communication: What does the team need to achieve through the communication?
- The most *appropriate* information: the most effective *message* – its format and delivery method.
- *Targeted* communications to meet the expectations and requirements of the stakeholder and the capacity and capability of the team.

Based on each stakeholder's unique engagement profile, a communication plan can be developed. The communication plan should contain:

- *Mutuality*: How the stakeholder is important to the activity and the stakeholder's *stake* and expectations.
- Categorization of influence (*upward, downward, outward, sideward, internal, and external*) and/or the categories of (Fassin, 2012). Stakeholders from each category will require messages presented in different content and format[14].
- Engagement profile preferably in graphical form: Level of support for the activity and receptiveness to information about the work AND target engagement: necessary levels of support and receptiveness.
- Strategies for delivering the message:
  - *Who* will deliver the message?
  - *What* the message will be: regular activity reports or special messages?
  - *How* it will be delivered: formal and/or informal, written and/or oral, technology of communication – e-mails, written memos, meetings?
  - *When*: how frequently it will be delivered, and over what time frame (where applicable)?
  - *Why*: the purpose for the communication: this is a function of mutuality – why the stakeholder is important for activity success, and what the stakeholder requires from the activity?
  - *Communication item*: the information that will be distributed – the content of the report or message.

Effective Communication

Regardless of how well the communication strategy and plan are crafted, other factors must be considered, such as development of messages that meet the specific needs of different individual or groups:

- The sponsor or other *upward* stakeholder may require concise summary information such as exception reports, with briefing data sufficient to be able to defend the activity.
- The *upward* stakeholder will also require early warning of potential negative events or actions that could embarrass that stakeholder. The key intent is to ensure that there are no *surprises*.
- Middle managers who supply resources need time frames, resource data, and reports on adherence to plans and effectiveness of resources provided, more comprehensive information.

- Staff working on the activity and other team members need detailed but focused information that will enable them to perform their activity roles effectively, such as Earned Value reports.
- Other staff need progress updates of information on how it will affect their own work roles.
- External stakeholders will also require targeted updates on aspects of the activity relevant to their interests in the activity, its deliverables, its impact, its progress.

Factors that Affect Communication Effectiveness

Other factors may act as barriers to effective communication. Awareness of these factors and their consequences may drive the timing and context of the communication activity. They will be described in more detail in the next section of the chapter.

- Personal reality: conscious and unconscious thought processes will influence how individuals receive and process any information they receive.
- Cultural differences: differences in communication requirements may be caused by cultural norms influencing the preferred style of presentation, content, and delivery of information. These differences may be national, generational, professional, and organizational (Zemke et al., 2013).
- Personality: personality differences may also dictate the *how* and *what* of effective communication. A senior manager with limited available time and a preference for summary information will have no patience for information delivered as a story, whereas a team member or a stakeholder with a different personality style may find the delivery of facts not interesting enough.
- Environmental and personal distractions will include effects of external issues such as pandemics and other global events; noise – either excessive environmental sound or an overwhelming load of useless and unusable data and information; lack of interest, fatigue, or personal issues. If either the sender or the receiver is known to 'have a bad day', or is feeling unhappy, it is better to postpone any face-to-face communication until another occasion.

Perception and 'Reality'

The answer to the question – *What makes us who we are and how do we operate in our social world?* – lies in a complex web of an individual's own 'reality' formed by unique experiences (and how the brain makes sense of those experiences), culture, whether national, professional, generational, and gender. The result is a unique assembly of characteristics that influences how that individual lives and works and relates to others. And within the work environment, the organizational culture affects the work and outcomes of the portfolio, programme or project and its stakeholders.

Researchers have long taken an interest in the different ways that individuals make sense of their surroundings – their 'world'. Weick (1995) developed the concept of sensemaking by which people strive to comprehend of their environment. Sensemaking refers to how we interpret an unfamiliar situation within the framework of previous experience, then incorporating the new data, or resolving the current issue, or adapting to a new environment[15]. Weick (1995) interpreted this process in one way, while neuroscientists have taken a completely different approach for how we construct our reality and how we learn and make sense of new situations[16].

Neuroscience explains how we deal with unfamiliar situations in a different way: when new information or situations are presented to us, the new data are compared with existing mental maps to find connections between new data and existing frameworks. If there are no connections, the brain will try to make the connections fit into the existing framework. New information or stimulation bombard the brain and continually forces it to take shortcuts. We have expectations about what we are going to read or experience, and therefore, we 'see' it in that frame – not necessarily what actually happened. Such approximation means that often we misunderstand or misinterpret what we observe.

For any individual, reality is understood through unique filters of experience, knowledge, and interests. Each person has constructed a different reality; he or she may describe the same scene in totally different ways[17]. The brain comprehends and interprets the world according to its own wiring, selecting and ignoring information depending on its the experience, knowledge, and interests that have created that unique reality.

Personality

Personality is a second factor to consider in engaging stakeholders. The term refers to an individual's distinct pattern of thoughts, motives, values, mindsets, attitudes, and behaviours. There are many typologies for categorizing personality; the best known of these typologies is the Myers-Briggs Indicator (MBTI).

The MBTI measures psychological preferences in how people perceive the world and make decisions (Kroeger & Thuesen, 1988). Often people will act in ways that cause others confusion or anger, because it seems wrong or unreasonable, but in fact is just different from how another individual might act. Kroeger & Thuesen (1988) state that when this happens it is not the problem of the person who seems to be wrong or unreasonable, it is the person who is feeling the anger! Their view is that it is important to try to understand why these others are acting the way they do and why it is causing such a reaction. They turn to a typology such as the MBTI to assist in this process.

The MBTI uses four pairs of alternative preferences which when combined provide 16 possible types of personalities[18]:

- Introversion (I) or Extraversion (E) – attitudes
- Sensing (S) and Intuition (N) – functions
- Thinking (T) and Feeling (F) – functions
- Judging (J) and Perception (P) – lifestyle

This analysis of personality attributes or behaviour preferences may assist understanding stakeholders' actions and reactions. As with any method or schema it is important to note that this schema can only provide guidance – never a definitive answer.

Culture

Personality and reality define characteristics of individuals and are not necessarily dependent on cultural influences – the individual's nationality, age, race, or gender. Hofstede et al. (2010) have defined culture as: 'software of the mind'. A person's culture (national, professional, and organizational) influences how messages will be sent and received (their communication style), which in turn influences how people from different backgrounds can work together harmoniously. Understanding how cultural background affects communication style reduces misunderstandings and helps build empathy.

Culture is learned – from parents, teachers, peers, and 'heroes' throughout childhood and well into adult life. Hofstede et al. (2010) have defined four ways to describe how culture manifests itself – symbols, heroes, rituals, and values. From an analysis of these four it is possible to analyse components of a culture and to provide a means to compare cultures.

- *Symbols are words, gestures, pictures, or objects that carry a particular meaning that is recognised as such only by those who share the culture.* In the context of delivering organizational value through relationships with stakeholders, the central concern of this chapter, the symbols will relate to the language, processes, and practices developed to frame the structure and the discipline of portfolios, programs, and projects within an organization.
- *Heroes are persons, alive or dead, real or imaginary, who possess characteristics that are highly prized in a culture and thus serve as models for behaviour.* In the context of this chapter, organizational heroes are often those in leadership roles such as a CEO who delivers organizational success through increasing shareholder value; management using their MBA qualifications to devise policy based on business case and bottom line; or the accountant who is charged with managing the business case or bottom line (Hofstede et al., 2010); or the heroic PMO manager or the portfolio, program or project manager who goes to extraordinary lengths to keep the work on track.
- *Rituals are collective activities that are technically superfluous to reach the desired end but that within a culture are considered socially essential. They are carried out for their own sake.* Rituals can range from how we pay respects to others, religious ceremonies, or business conferences (a way of reinforcing group identity). *Rituals include discourse, the way that language is used in text and talk, in daily interaction and in communicating beliefs.* Many meetings and progress reports are ritualistic observances that perform no real business function.
- *Values are broad tendencies to prefer certain states of affairs over others.* Because they are acquired early, many values remain unconscious to those who hold them. The value system is central to culture and is best understood through understanding pairings such as: good/ evil; dirty/clean; dangerous/safe; or abnormal/normal; paradoxical/ logical; irrational/rational (Hofstede et al., 2010).

In the organizational project environment, values are often expressed as *on time, within budget, and to scope.* This approach may not necessarily be the best approach for delivering value to the organization; it may even impose undue stress on those who must deliver the outcomes constrained by limited resources and funding, and the fluctuating interest of important stakeholders.

Cultural diversity within a project team can take the following forms:

- Generational and gender – a team may contain representatives from as many as four different generational groups: baby boomers, Gen X, Y, or Z. Generational differences may cause misunderstandings based on communication preferences, attitudes to work, and even language.
- Industrial or professional – Managers; professionals (engineers, accountants, and teachers); and operational personnel. They will have different communication styles, language, and approaches to work.
- National – consider a mix of Asian, Anglo-American, and Latino cultures – here also there will be different communication styles, language, and approaches to work.
- Organizational - Corporations, Government departments, and Universities will all have different structures and focus.

Hofstede et al. (2010) developed typologies of culture from research carried out for IBM in the 1980s and updated in 2010. Initially five dimensions were defined and with recent collaborations with other researchers a sixth was added[19]:

- Power distance (weak/strong) (PDI): an indicator of dependence relationships in a country.
- Collectivism/individualism (IDV): Individualism defines societies in which the ties between individuals are loose: everyone is expected to look after himself or herself and his or her immediate family. Collectivism defines societies in which people from birth onward are integrated into strong, cohesive in-groups, which continue to protect people throughout their lifetime in exchange for unquestioning loyalty.
- Femininity/masculinity (MAS): masculine society defines gender roles as distinct. Men are supposed to be focused on material success, and women are supposed to be more concerned with quality of life. In a feminine society, gender roles overlap; both men and women are supposed to be modest, tender, and concerned with the quality of life.

- Uncertainty avoidance (UAI): members of a strong UAI culture feel threatened by ambiguous in unknown situations. There is a need for predictability in the form of written and unwritten rules.

Long-term/short-term orientation (LTO): LTO stands for the fostering of virtues oriented toward future rewards, in particular perseverance and thrift; short-term orientation (STO) stands for the fostering of virtues related to the past and present in particular respect for tradition, preservation of 'face', and fulfilling social obligations.

- Indulgent/restrained: This dimension has as its focus happiness – a universally cherished goal. There are two main aspects: evaluation of one's life and description of one's feelings.

Generational Culture

From time immemorial, older generations have complained about the transgressions and lack of respect of the younger generations; complaints about the loss of respect of the younger generation were found in Egyptian scrolls 2000 BC (Hofstede et al., 2010). In a multi-generational team, there can be many areas of potential conflict and misunderstanding:

- Symbols, heroes, and rituals,
- Differences in values and points of view,
- Ways of working and thinking, and
- Talking and communicating.

To build strong relationships between the portfolio, program, or project and its stakeholders, it is important to understand these generational differences to bridge the inevitable gaps. The focus on generational studies using cohorts labelled as Boomers, GenX, Millennials, etc., is useful in building such relationships. It is also important to recognise that as members of these cohorts grow older, they will adapt to their different life circumstances – careers, marriage, parenthood, retirement. Different life stages will add further complexity to an understanding of how these different generational cohorts operate and how the other generational groups view them[20].

Professional Culture

Professional cultures are also different communities within the team or within organizations. Schein (1996) identified three distinct cultures in manufacturing organizations that translate quite seamlessly into organizations of the twenty-first century:

- Operators – Exist within the part of the organization that builds the product or delivers the service. Their structure and values are unique to the organization or at least to their industry. Their culture and norms are built on trust and teamwork.
- Technical specialists – Designers and implementers of technology. These categories include project managers, engineers, and hardware specialists.
- Executives – Fiscal responsibility. They favour command and control systems and management techniques based on command and control.

Communication between these different groups can be improved through the efforts each group makes to understand the values, symbols, and rituals of the other groups.

Gender

The previous sections on perception, personality, and culture have focused on factors that cause individuals to think and act in the way that they do. It illustrates how each individual is unique. There is one more point of difference that needs to be considered within this chapter, the difference of gender.

The social context – national, community, family – we grow up in influences who we are, how we think, and what we do. All our social expectations and stereotypes are formed at an early age. In Western culture gender stereotypes fit, to a greater or lesser extent, within the framework of the following:

Female traits (Fine, 2010):

- Communal personality traits
- Compassionate, dependent, interpersonally sensitive, and nurturing
- To serve the needs of others

The male traits:

- Agentic personality traits[21]
- Aggressive, leader, ambitious, analytical, competitive, dominant, independent, and individualistic
- To bend the world to your command and earn a wage for it[22]

A person grows and changes in response to social environment. Each person develops a 'Wardrobe of Self' (Fine, 2010) to match all the social identities one person can assume[23].

The gender stereotypes that provide the clothing for the Wardrobe of Self are reinforced by society. In masculine countries, such as United States, Australia and the United Kingdom, boys are socialised toward assertiveness, ambition, and competition. In these same countries until recently, girls were polarised between some who wanted a career and many who did not. Currently, in many countries women work for economic reasons or for personal fulfilment; she is still expected to balance the responsibilities of caring for family and meeting the requirements of her career.

Discourse: The Sharing of Information

Communication with stakeholders requires consideration of all the differences in culture already outlined, with some final words about differences in the way that men and women transmit and choose to interpret information. Tannen (2013) describes these differences as:

- *Report talk*: – the way that men communicate both formally and informally, transferring information to establish and maintain status that displays their abilities and knowledge.
- *Rapport talk*: the way that women communicate both formally and informally to build and maintain connections, first validating the relationship to build rapport and then dealing with any business.

Neither of these ways of communicating is necessarily superior to the other – this is just how men and women have been socialized. But it also explains why there can be misunderstandings in both formal and informal conversations, where men try to *fix* a problem by giving advice, and women want to talk about the problem without necessary needing the advice men bestow. These misunderstandings will also explain the impression that many male managers have of the linguistic styles of their female colleagues. For example, women ask more questions, usually for clarification or deeper understanding: this has been interpreted by male managers as not knowing enough (Tannen, 1995).[24]

Organizational Culture

The final aspect of culture is the specific culture of an organization. When companies are part of international corporations, their planning and control systems will be influenced by the national culture specific to the country in which this branch of the company practices, even though headquarters will attempt to influence decision-making, processes, and controls. Different organizations will display different characteristics, depending on their structure and mission: corporations (for profit); Not for Profit, such as charities; and Government Departments or agencies. Within these characteristics will be other distinguishing features based on:

- Risk tolerance – are they risk avoiding or risk seeking?
- Charter – are they entrepreneurial or public-service oriented?
- Who benefits – shareholders? Selected groups of society? Or the public at large?
- Product orientation – manufacturing, product sales, service providers, or a mixture?
- National, regional, or multinational?

The culture of the organization will be formed from the mix of features: in turn the culture of the organization will influence the style of management within the organization. Part of the mental software of individuals working within the organization consists of their ideas about what an organization should be like; *power distance* and *uncertainty avoidance* affect thinking about organizations (Hofstede et al., 2010). Understanding these dimensions requires answering two questions:

- Who has the power to decide what (*power distance*)?
- What rules or procedures will be followed to attain the desired ends (*uncertainty avoidance*)?

CONCLUSION

Only when the needs (expectations) of each key stakeholder and the stake or stakes he or she may have in the outcome are known and understood is it really possible to begin to understand their drivers or business needs. These expectations may be overly ambitious or unrealistic. With an understanding of the cultural background of key stakeholders, messages can be crafted, and information offered to inform stakeholders about what can be achieved, and how these achievements converge with their expectations.

198 • *Portfolio Management*

The needs of important stakeholders cannot be assumed and may not even be related to the work itself. But this understanding of needs and expectations is crucial. It is a fundamental starting point for any campaign to reach agreement on alternative objectives and the management of the perception of the stakeholder *but also* the perceptions of the stakeholders of that stakeholder.

The focus of this chapter has been on describing what can happen in the *Zone of Uncertainty*, the area between an organization's strategic vision and the projects created to deliver that vision. It describes the murky, complicated, and complex activity to achieve the vision. It also describes the difference between the trust in linear processes regulated by controls, and the messy reality of its implementation. There are three major causes of the complexity that creates the Zone: technological complexity, the complexity of the work that results from seeking to achieve balance in the portfolio, and the *stakeholder factor*. The *stakeholder factor* describes the complexity of human relationships and human activity as the basic building block in the work of organizations. Improvement in stakeholder relationships will reduce the messiness and confusion that everyone experiences within the Zone and provide guidance for the PMO, portfolio, program, or project manager. Stakeholder relationship management is not easy; it is also very time-consuming; it cannot be reduced to templates or standard e-mails. To deliver value to the organization effectively through portfolios, programs, and projects, analysis of the stakeholder community and its engagement requires exchange of information–communication. Successful communication requires not only careful planning and thoughtful implementation, but it also requires understanding more about each individual or group that makes up the stakeholder community, through an understanding of their cultural background and their mental maps – how they make sense of their world and how they learn and act. Ultimately the stakeholder factor and successful actions to engage important stakeholders and meet their expectations can add clarity to the murky depths of the Zone. Engagement is complex and unpredictable and requires vigilance; reliance solely on policy and controls will lead to calamity and disappointment.

NOTES

1 Their attention is now on other visions and strategies, believing that the process will automatically proceed without any more senior management intervention.
2 Complicated = a large number of interconnected and interdependent parts.

3 Project work is complex if it consists of many interdependent parts each of which can change in ways that are not totally predictable and which can then have unpredictable impacts on other elements that are themselves capable of change (Cooke-Davies, Crawford, Patton, Stevens, & Williams, 2011).

4 In interviews with managers in organizations in the Spanish-speaking countries in South America, I discovered that 'stakeholder' translated has many meanings, often focusing on just one attribute rather that the more inclusive definition becoming more accepted in the English-speaking world. This led to my informal enquiry of my international colleagues regarding how 'stakeholder' was translated.

5 I presented at conferences in Brazil in 2009 and 2011– this is a consistent theme of all discussions with project managers and Project Management Office (PMO) practitioners in that country.

6 I had the opportunity to speak to a group of project managers in Japan in January 2013 and ask the question of how 'stakeholders' were translated in Japanese at that time.

7 From e-mail correspondence between one of the authors and Bob Youker, previously working for the World Bank, now retired.

8 The concepts defined in this chapter and the methodology applies to ALL activities that an organization approves, resources, and funds to achieve its strategies and goals.

9 The definition of *power* will only depend on the joint agreement of the team making the assessment of the ability to cause change: it does not depend on any other definition.

10 *Proximity* simply defines how stakeholders are involved in the work of the project or activity.

11 Reading the *Stakeholder Circle* map of stakeholders is best in colour. For examples of the outputs of *Stakeholder Circle*, go to: https://mosaicprojects.com.au/shop-SHC-tools.php

12 Colours are not shown here: For a full colour representation, refer to https://mosaicprojects.com.au/shop-SHC-tools.php or Bourne (2009).

13 These elements are also aspects of any individual's cultural background which will be discussed later in this chapter.

14 For example, a stakeholder from the *upward* category will probably only need a summary in the format that is most familiar – financial professions probably require spreadsheet information, and human resource managers may prefer graphics. Communication to the team, however, will be more effective in detail with clear instructions on what has to be done and how will do it.

15 To illustrate this concept, Weick (1995, p. 55) relates the story of a small military unit sent on a training mission into the Swiss Alps and who became lost in a snowstorm. One of them had a map, and with the assistance of that map they planned their journey back to their base. When the storm subsided, they began their journey back to base. On that journey, they did not always find the landmarks that the map showed, but with the help of residents of the villages they passed through they eventually found their way back to base, tired, hungry, and cold. That was when they discovered that the map was a map of the Pyrenees and not the Alps! The map was not the blueprint, it was only the artefact that helped them get started; the rest of the journey was facilitated by cues from the environment, incorporating new information and acting with purpose.

16 There are obviously many other theories contributing to an understanding of how we construct reality. It has long been a question that philosophers have grappled with – reality and the relationships between the mind and reality through the means of language and culture.

17 Horowitz (2013) describes what happened when she turned a daily walk around the block with her dog into an exercise of perception. She invited people from different professions to walk with her and describe what they saw. Each one of them drew her attention to different aspects of the same pathways she had walked on many times before: psychiatrist, economist, her 19-month-old son, an architect, and eight others. They all described aspects of that block that she could never have imagined; it was outside her experience.

18 For most effective results, it makes sense to use the MBTI as used by groups who have received the appropriate training in applying the instrument and analysing results. However, for a quick assessment to get a feel for the MBTI process, go to: http://www.personalitypathways.com/type_inventory.html

19 The website: https://geerthofstede.com/culture-geert-hofstede-gert-jan-hofstede/6d-model-of-national-culture/ provides essential information about profiles and how to interpret them

20 For example, *twenty-somethings* are ideological and believe that the older generations are cynical and complacent; at 30 most individuals will have met a life-partner and may even have begun to raise a family – this will change their world view. At 40, individuals will begin to recognise that many of their dreams may now never be fulfilled – this can lead to the phenomena of changing jobs of making life style changes. At 50 people have more money and fewer expenses and can indulge in things that they could not afford in their youth, such as fast cars or motor bikes. These opinions are based on my own observations of Western individuals and groups and some conversations with people in these age groups.

21 The capacity to exercise control over the nature and quality of one's life: the capacity to act in the world (National Centre for Biotechnology Information).

22 This is the case for white, middle-class heterosexual men (Fine, 2010).

23 My wardrobe is: Melbourne resident, teacher, grandmother, woman, university professor, writer, Baby Boomer. Depending on which identity I need to 'wear' I will have different approaches, perhaps use different language and tone, I will socialize in different ways. Who I am is sensitive to the social context at that moment.

24 And there is the story of how women are willing to ask for directions, whereas men are reluctant to do so.

REFERENCES

Bourne, L. (2012). *Stakeholder relationship management: A maturity model for organisational implementation* (Revised Edition). Farnham, UK: Gower.

Bourne, L. (2014). *Making projects work: Effective stakeholder management and communication*. Boca Raton, FL: CRC Press.

Cooke-Davies, T., Crawford, L., Patton, J., Stevens, C., & Williams, T. (Eds.). (2011). *Aspects of complexity: Managing projects in a complex world*. Newtown Square, Pennsylvania: Project Management Institute.

Dinsmore, P. C. (1999). *Winning in business with enterprise project management*. New York: AMA Publication.

Fassin, Y. (2012). Stakeholder management, reciprocity, and stakeholder responsibility. *Journal of Business Ethics*, 109, 83–96.

Fine, C. (2010). *Delusions of gender: The real science behind sex differences.* London: Icon Books.

Hofstede, G., Hofstede, G. J., & Minkov, M. (2010). *Cultures and organizations: Software of the mind. Intercultural cooperation and its importance for survival.* New York: McGraw-Hill.

Horowitz, A. (2013). *On looking: About everything there is to see.* London: Simon & Schuster.

Kroeger, O., & Thuesen, J. (1988). *Type talk: The 16 personality types that determine how we live, love, and work.* New York: Dell Publishing.

Mitchell, R. K., Agle, B. R., & Wood, D. J. (1997). Toward a theory of stakeholder identification and salience: Defining the principle of who and what really counts. *Academy of Management Review, 22*(4), 853–888.

Project Management Institute (PMI) (2021). *Project management standard.* Newtown Square, Pennsylvania: PMI.

Schein, E. H. (1996). Three cultures of management: The key to organizational learning. *Sloan Management Review, Fall, 1996,* 9–20.

Tannen, D. (1995). The power of talk: Who gets heard and why. *Harvard Business Review, 'on communicating effectively'.* September, 1995.

Tannen, D. (2013). *You just don't understand: Women and men in conversation.* New York: HarperCollins.

Weick, K. E. (1995). *Sensemaking in organizations.* Thousand Oaks, CA: Sage.

Zemke, R., Raines, C., & Filipczak, C. (2013). *Generations at work: Managing the clash of boomers, gen xers, and gen yers in the workplace.* New York: Amacom – American Management Association.

11

Strategic Alignment for Agile Portfolios

Yvan Petit, Alejandro Romero-Torres, and Julie Delisle

INTRODUCTION

Most of the other chapters of this book cover project portfolio management in a "traditional" context in organizations where projects are launched and managed under the umbrella of portfolio management tools and techniques. The underlying condition is that projects can be evaluated, prioritized, and selected based on the traditional project planning constraints *schedule, cost and scope*. Traditional approach considers long-time periods for planning; in contrast with agile approaches that triggers iterative and short time period planning. Over the last two to three decades traditional approach has demonstrated its benefits primarily to ensure that projects in the portfolio are properly aligned with the organization's strategy, thus ensuring that the organization invests resources in the right project (Dinsmore & Cooke-Davies, 2006). Standards have been developed to document best practices (Axelos, 2011; ISO, 2015; Project Management Institute, 2017).

The advent of agile approaches (Agile Alliance, 2001) has challenged the way that software development projects are managed. In many instances the development cycle is no longer defined as a project with a beginning and end but might resemble a production line with fixed capacity and continuous production of product releases (Fitzgerald & Stol, 2017; Narayan, 2015). Agile was originally designed with a single project team in mind but the benefits of agile have enticed organizations to agile enterprise-wide (Dikert et al., 2016; Rigby et al., 2018). Inevitably this has also challenged the way that organizations manage portfolios with multiple agile initiatives or with initiatives including multiple teams (Hobbs & Petit, 2017b, 2017a), using a mixture of traditional, agile or hybrid methodologies to execute projects (Gemino et al., 2021).

Various frameworks and models have been developed to address the problem of scaling agile (Ambler & Lines, 2016; Larman & Vodde, 2014; Scaled Agile, 2020; Vaidya, 2014). Organizations have been adopting these frameworks and models to harvest the benefits of agile and have seen improvements in the overall performance of the organization (Dikert et al., 2016). A challenge created in a scaled agile environment is how projects or initiatives should be managed in such a way that they contribute to the organizational vision and strategies.

This chapter attempts to describe how organizations ensure that agile initiatives align to the organization strategy. After reviewing and comparing the traditional project portfolio models to agile portfolio management, the portfolio element of scaled agile models is explored with a particular focus on the Scale Agile Framework (SAFe), which is the most widespread (Digital.ai, 2021). Although research on agile project portfolio management is still scarce and requires further investigation (Petit & Marnewick, 2021; Stettina & Hörz, 2015; Sweetman & Conboy, 2018), an overview will be provided, and some elements of what organizations do in practice will be presented through a concrete case in this chapter. Because in fine, value is what we aim for with portfolio management (no matter the approach), we organized this chapter using four key goals and one overarching mechanism rather than processes to explore and compare traditional and agile approach to portfolio management. Although we present each one in sequence, they are not permeable boxes, and some processes could be related to more than one goal.

THEORETICAL BACKGROUND

Strategic Alignment of Projects through Project Portfolio Management

A project portfolio is defined by the PMI as a "collection of projects, programs, subsidiary portfolios and operations managed as a group to achieve strategic objectives" (Project Management Institute, 2017, p. 3) while ISO

21504 defines it as: "a collection of portfolio components grouped together to facilitate their management to meet, in whole, or in part, an organization's strategic objectives" (ISO, 2015, p. 1). Project portfolio management per se is then the "centralized management of one or more project portfolios to achieve strategic objectives" (Project Management Institute, 2017, p. 5).

Our literature review (Ahmad et al., 2017; Beringer et al., 2013; Cooper et al., 2001; Martinsuo & Lehtonen, 2007; Pennypacker & Dye, 2002; Stettina & Hörz, 2015) brought forward four main goals for Portfolio:

Goal 1 - Strategic direction: Aims to ensure that the final portfolio of projects reflects the business's strategy, that the breakdown of spending across projects, areas, markets, etc., mirrors the business's strategy, and that all projects are on strategy (Pennypacker & Dye, 2002, p. 196–197). There is unanimity in the literature on Goal 1 and it is widely accepted that projects are used to implement organizational strategies (Cooke-Davies et al., 2009; Hermano & Martin-Cruz, 2016). The last ten years have also seen an increase in the number of publications on project alignment with business strategy (Dinsmore & Cooke-Davies, 2006; Garfein, 2005; Lanka, 2007; Lan-Ying & Yong-Dong, 2007; Milosevic & Srivannaboon, 2006; Shenhar et al., 2007), including the first edition of this book (Levin & Wyzalek, 2014).

One of the important aspects of the strategic alignment of programs and projects is the strategic value that is created and released through this alignment. Since projects enable the successful delivery of long-term strategies, they should find ways to unleash strategic value through the alignment of programs and projects (Martinsuo et al., 2012; Project Management Institute, 2017) even when facing uncertainty in dynamic environments (Killen et al., 2012, 2013; Petit & Hobbs, 2010, 2012).

Shenhar et al. (2007) view alignment of project management and business strategy as "an internal collaborative state where project activities continually support the achievement of enterprise strategic goals" (p. 7). In the same way, Morris and Jamieson (2005) observe that PPM is primarily about selecting and prioritizing projects, not managing them.

Goal 2 – Value Maximization: Aims to allocate resources to maximize the value of the portfolio in terms of some business objectives, such as profitability. One way to achieve that is to determine the values of projects to the business, and rank projects according to this value until there are no more resources. Prioritization is the process of ranking the selected components based on evaluation scores. This is a process which has received little attention in the literature in comparison to the selection and balancing techniques. The prioritization techniques are often simple; for example, weighted ranking, scoring techniques, or expert judgment.

Cooper (1997a, 1997b, 1998, 2001) has developed some of the earlier forms of PPM processes and connected them to the stage-gate technique, a decision process for individual projects (Cooper, 2008; Cooper et al., 2002a, 2002b).

A number of scoring models and financial techniques such as net present value, dynamic rank ordered list, expected commercial value, real options, checklists and productivity index have been used in the industry and surveyed by Rad and Levin (2007). Additional techniques include scenario planning (Dye, 2002), what-if analysis (Benko & McFarlan, 2003), decision trees (Gustafsson & Salo, 2005), scoring techniques (De Piante Henriksen & Jensen Traynor, 1999), and portfolio management indices (Rad & Levin, 2005).

Goal 3 – Portfolio Balance: Aims to achieve a desired balance of projects in terms of a few parameters: long-term projects versus short-term ones; high-risk versus sure bets; and across various markets, technologies, and project types.

According to PMI,

> [T]he objective of portfolio risk management is to meet the value proposition of the portfolio while aligning to an agreed-upon confidence level and/or portfolio-level risk threshold through a balancing process of both threats and opportunities. Risk balancing thus contributes to maximizing the probability that the portfolio will support strategic objectives within the value proposition constraints.
>
> (Project Management Institute, 2017, p. 86)

Different graphical representations of the projects, for example using bubble charts are now used. This modern adaptation of the Boston Consulting Group matrix maps the different projects according to two axes (for example, benefits versus risks). It uses the size of a circle to represent the cost of the projects, the color of the circle to represent another variable, such as the timing, and shading to represent yet another variable such as the product line. Balancing processes ensure that the organizations' constraints are taken into consideration and according to (2003):

> [E]very organization has two constraints that limit how many projects can be active at any point in time. One is the amount of money the organization has or is willing to invest in change. The other is the organization's strategic resources – the one most in demand across many projects or the most heavily loaded resource across most projects. This determines how many projects can be active at any point in time.
>
> (p. 211)

This is the process in which the best mix of projects is identified in order to achieve the organization's strategic goals. The best mix might actually not only include the projects with the highest values or lowest risks. Unbalanced portfolios might result in too many projects of a certain type at the expense of another type resulting in increasing risk exposure.

PMI suggests three principles that can be related to goal 3 and considered central to risk management processes at the portfolio level: maximize portfolio value while balancing risks; foster a culture that embraces change and risk; and navigate complexity to enable successful outcomes.

Goal 4 – Portfolio performance: According to Müller, Martinsuo and Blomquist (2008), portfolio performance measures in practice are multidimensional and include the project, portfolio, and organizational level. In their quantitative research, they identified three portfolio management performance measures: achieving results, achieving purpose, and balancing priorities. Since the latter is already covered under Goal 3, the focus of portfolio performance in this chapter comprises only the first two achieving results and purpose.

This corresponds to the periodic assessment of the portfolio to determine its performance along key indicators and metrics such as evolution towards results, spending, risks and dependencies. This is a management's opportunity to gather the necessary information about the portfolio to be able to re-align the projects, if necessary. According to the PMI Standard for Portfolio Management (2017):

> Monitor and control is one of the critical supportive activities for monitoring portfolio performance and recommending changes to the portfolio component mix and portfolio component performance and compliance with organisational standards. The purpose of monitor and control is to understand when changes need to be made to the portfolio or to the portfolio management processes. This process includes execution, documentation, and communication of the decisions and the resulting actions taken.
>
> (p. 26)

This might result in the addition, reprioritization, or exclusion of some projects in addition to new directives and rebalancing of the portfolio. McDonough and Spital (2003) found that portfolios being reviewed quarterly performed better than those reviewed semi-annually. Although they hypothesize that the optimal frequency of portfolio review might depend upon the type of projects and the dynamism of the industry, they did not have sufficient data to test these relationships. They also observe that project termination is often a more difficult managerial decision than project approval, which corroborates similar findings by Cooper, Edgett and Kleinschmidt (2001) and Royer (2003).

OVERARCHING MECHANISM: PORTFOLIO GOVERNANCE

PMI defines portfolio governance as "the framework, functions, and processes that guide portfolio management activities in order to optimize investments to meet organizational strategic and operational goals" (Project Management Institute, 2016, p. 3). They further describe portfolio governance using four governance domains:

- **Governance Alignment**: Functions and processes to create and maintain an integrated governance framework.
- **Governance Risk**: Functions and processes to identify and resolve threats and opportunities to ensure balance of risk and reward.

- **Governance Performance**: Functions and processes to ensure measurement and evaluation of KPIs against parameters and realization of business value.
- **Governance Communications**: Functions and processes to disseminate information, engage stakeholders, and ensure organizational change.

As we can see, this goal is overarching and transversal to the previous ones presented, as governance has to do with strategic direction, value maximization, balance and portfolio performance. Standard processes have been developed to support this alignment such as identification, categorization, evaluation, selection, and prioritization. Some authors have also proposed the use of decision points called Stage-Gate (Cooper et al., 2002a, 2002b) which could either be used for traditional projects but also combined with agile practices (Conforto & Amaral, 2016)

At last, uncertainty related to scope of projects along with dynamic environments lead to the introduction of agile and scaling agile within organizations. In that sense the iterative nature of agile methods with frequent re-evaluation of project results impacts current portfolio management practice (Stettina & Hörz, 2015), which will be covered in the next section.

Agile Portfolio Management

Not only agile projects do not necessarily fit traditional portfolio management processes seen earlier, but their iterative nature brings new challenges to current practices (Stettina & Hörz, 2015). An agile portfolio can be defined as a "project portfolios that to a relevant share constitute of agile projects (i.e., single projects managed according to agile practices)" (Kaufmann et al., 2020, p. 430). As such, an agile portfolio will be composed of many small projects building on progressive and just in time planning and a greater presence of changes, which all have impacts on the way a portfolio is managed, and strategic alignment is ensured.

Scaling agile means extending agile practices across different teams of an organization. For example, large-scale agile has been defined as representing 50 or more people or at least six teams (Dikert et al., 2016). The challenges to scale agile practices to the portfolio level have been recognized (e.g. Kaufmann et al., 2020). For example, the greater extent of interactions within and across projects and actors increases portfolio level complexities (Puthenpurackal Chakko et al., 2021). This greater extent of interactions and constant changes increase the need for coordination and adaptiveness (Scheerer, et al. 2014), with an impact at the portfolio level, since a traditional approach is no longer appropriate (Sweetman & Conboy, 2018).

Challenges are bigger for larger organizations in which frequent feedback loops, iterative reviews and close customer contact, keys to agile practices, are more challenging to accomplish due to routines and structures (Stettina & Hörz, 2015). As reported by these authors, challenges stem from the difficulty to align business needs and strategy, pursuing prioritization, resource allocation and governance (Rautiainen et al., 2011; Thomas & Baker, 2008), and synchronizing dependencies (Hodgkins & Hohmann, 2007; Kalliney, 2009).

Despite differences between traditional and agile portfolio management, we assume that the goals stay the same, that is: ensuring strategic direction (goal 1), maximizing value (goal 2), balancing the portfolio (goal 3), ensuring portfolio performance (goal 4). All rely onqqqq portfolio governance as an overarching mechanism to achieve these goals. Portfolio governance ensures that functions and processes are put in place to reach these goals. However, the way these goals are carried out is much different for agile portfolio management, which will be further explored in this chapter.

Scaled Agile Frameworks

Agile approaches have been developed for single team development. However, based on the success of these approaches, many organizations have begun to implement agile at scale, i.e. for large development initiatives or for a large number of initiatives (Alqudah & Razali, 2016; Dingsøyr et al., 2014; Hobbs & Petit, 2017b, 2017a). In the last decade, numerous frameworks have been proposed to structure the scaling of agile and assist organizations in their implementation of agile at scale.

Nielsen (2021) summarizes the core concepts of 10 scaled agile frameworks with a special attention to the supporting elements at the program and portfolio level. Uludag et al. (2017) surveyed 20 frameworks and assessed their level of maturity using parameters such as the number of academic contributions, the number of case studies, the available documentation that could be found, the training courses and certifications offered by the organization, and the content of the community forums and blogs. Three frameworks stand out as most mature: Large Scale Scrum (LeSS) (Larman & Vodde, 2014), Disciplined Agile (formerly DAD) (Ambler, 2017, 2020), and Scaled Agile Framework (SAFe) (Leffingwell, 2017; Scaled Agile, 2020).

Annual surveys by digital.ai (formerly VersionOne) have confirmed over the last few years that SAFe, Scrum of Scrum, DA and LeSS are used by over 50% of the organization implementing agile at Scale (Digital.ai, 2021). This being said, agile portfolio in practice is frequently initiated in projects and not following a specific framework portfolio framework (Stettina & Hörz, 2015), as many companies prefer to develop their own customized method (Edison et al., 2021).

In addition to the maturity and market share of the frameworks, an important aspect of this book chapter is to what extent the frameworks support the portfolio goals presented in the section "Strategic Alignment of Projects through Project Portfolio Management" of this chapter. Grundler & Westner (2019) researched the two following questions: What are the conceptual differences between scaling agile frameworks? and How do scaling agile frameworks conform to the objectives of traditional PPM? For the second question, they compared four of the main frameworks using the three portfolio goals defined by Beringer et al. (2013): Portfolio Structuring, Portfolio Steeering and Resource Management. They reached the conclusion that "SAFe offers the most complete and detailed framework, seeming to be the best fit for large organizations which require detailed descriptions of processes and roles for their program and portfolio structures" (p. 61).

In sum, although many organizations do not follow a unique framework as exposed earlier, SAFe is recognised as the most popular scaling method and is on the rise, which explains why we chose to focus on this framework in this chapter. The following section describes how main elements of SAFe are supporting the achievement of the portfolio goals. This is then compared with observations made in a deployment of SAFe in a large South-African bank.

SCALING FRAMEWORK SAFe

The scaling agile framework SAFe is a commitment to lean and agile values and principles in business organization. This methodology brings the ability to change and adapt business organization and strategy to deliver in short periods of time, by gaining competitiveness and prospering in the market (Rigby et al., 2018). To apply business agility, there are nine lean-agile principles (see Table 11.1) that promote continuous change in the

TABLE 11.1

Lean-Agile Principles for SAFe (Scaled Agile, 2020)

Principle	Description	SAFe Portfolio Component
Take an economic vision	Develop solutions in shortest sustainable development cycle by delivering organizational and/or customer value	Strategic themes and strategic portfolio review and participatory budget
Apply system thinking	Consider complexity and interdependencies (people and processes) to optimize organizational systems	Value streams
Assume variability and preserve options	Maintain multiple design options for longer periods to generate flexibility and better economic results.	Value streams and epics identification
Build incrementally with fast, integrated learning cycles	Develop solutions incrementally based on a series of short iterations to enable customer feedback, reduce risk and alternative actions.	Portfolio synchronization
Base milestone on objective evaluation or working systems	Evaluate progress with integration points to provide objective hits to evaluate the solution throughout the development life cycle.	Portfolio Kanban and portfolio synchronization
Visualize and limit WIP, reduce batch size, and manage queue lengths	Achieve a state of continuous flow, where the new capabilities of the solution change quickly and visibly from a conceptual level to reality.	Portfolio Kanban, epics, and enablers
Apply cadence, synchronize with cross-domain planning	Create predictability and provides a rhythm for development to operate effectively in uncertain environments.	Portfolio Kanban and portfolio roadmap and portfolio synchronization with program increment
Unlock the intrinsic motivation of knowledge workers	Provide people autonomy and purpose, while minimizing restrictions, to lead to higher levels of employee commitment, which translates into better results for customers and the organization.	Collaboration enablers, such as participatory budget
Decentralize decision-making	Enable rapid and decentralized decision-making process for regular decisions and enable centralization for strategical ones.	Strategic portfolio review, portfolio synchronization and participatory budget
Organize around value	Map organizational flow from customer requirements to value delivery to enable collaboration among organizational teams	Value streams and ARTs

organization, but at the same time, they sustain business operations and growth by adding value without failing in the attempt (Laanti & Kangas, 2015).

Based on the principles described above, SAFe focuses on implementing agility through the entire organization, and not only in development teams. Therefore, this framework includes three specific levels: team, program, and portfolio. In a simple way, at the team level, SAFe defines how each team that participates in the development of solutions is articulated. In this level, Scrum or Kanban techniques are proposed. At the program level, the organization and the objectives pursued are outlined even more. In this case, SAFe includes tools and techniques to define what work will be carried out by each team based on a chain of command and results, known as agile release train (ART). In this level, each team is involved in the planning and aware of activities and iterations for all the teams belonging to a same agile release train (or program). Finally, at the portfolio level, SAFe defines and promotes what contributes to creating value in the organization by enabling a systematic and economic view of the organization.

In the following sub-sections, we described how SAFe framework defines portfolio management. As shown in the Table 11.2, we described practices to enable portfolio goals and then practices related to the portfolio governance.

TABLE 11.2

SAFe Portfolio Management Goals and Governance

Portfolio Management Goals		Artifacts	Specific Ceremonies
Portfolio governance	Strategic direction	Strategic themes and portfolio vision	Strategic portfolio review (every three months) and portfolio synchronization (every month)
	Value maximization	Value stream, epics, enablers, agile release train and guardrails	Portfolio synchronization (every month)
	Portfolio balance	Funding value streams and planning by horizon	Participatory budget (every six months)
	Portfolio performance	Portfolio Kanban and portfolio roadmap	Strategic portfolio review, portfolio synchronization and participatory budget

How Are Portfolio Goals Achieved in SAFe?

In a SAFe organization, *strategic direction* (goal 1) is operationalized by collaboratively defining *strategic themes*. A strategic theme represents a specific and differentiated business intent and communicates aspects of the organization's strategic objective to the portfolio level (Scaled Agile, 2020). It defines the gap between the current state and the desired future state for the organization. Mostly, several *strategic themes* are defined to enable portfolio vision and alignment. Strategic themes are defined collaboratively by business drivers, also called *epic owners* (mostly business directors) who are supported by *enterprise architects and lean portfolio managers* to achieve collaboration and comply with portfolio governance (see section below on "Overarching Mechanisms for Portfolio: Governance"). They work all together to define a shared understanding of the organizational strategy, pursuit common goals and discuss across administrative units. By defining *strategic themes* actors responsible for the portfolio, including business drivers, can make collaborative decisions and ensure portfolio alignment to the organizational strategy.

In function of the strategic themes, business drivers ensure *value maximization* (goal 2) by distributing funds from the organization's budget to different *value streams*. A *value stream* is a set of interdependent actions that an organization executes to add value for its customers from their initial request through the value deliverable to customers. Mostly, organizational *value streams* correspond to business products and/or services that are required by customers and with specific strategic objectives to fulfill. Establishing value streams breaks silos since they enable analyzing how business units collaborate all together to satisfy customers' needs. With value streams, business drives can identify potential opportunities to accomplish strategic themes and then, they can define initiatives to undertake for accomplishing opportunities, also known as *epics*[1]. These initiatives include, for instance, technological or organizational investments which need to be carried out, or complex platform improvements necessary to support such investments. To communicate *epic* propositions, business drivers create *Epic Hypothesis Statements* where they describe the *Minimum Value Product*[2] concept for enabling the *epic*, its business outcomes, and its hypotheses. Business drivers also estimate human resource and asset requirements for realizing each epic and distribute them in one or multiples *Agile Release Train* (ART)[3].

Once *epics* and *ARTs* have been defined, *portfolio balance* (goal 3) is enabled by evaluating and prioritizing *epics* through the means of lean-agile budgeting (Scaled Agile, 2020). Lean-agile budget considers three main components: i) funding value streams and not projects by assigning a specific budget to value stream and by adopting funding guidelines, also known as *portfolio guardrails*, ii) distributing *epics* in four horizons: a year period for decommissioning end-life solutions (horizon 0) and improving core business (horizon 1), one to two years for next-generation products or services (horizon 2), and more than three years for exploring new opportunities (horizon 3); and iii) participatory budgeting where organizational funding is distributing equally to the business drivers who should negotiate to prioritize specific value streams These three lean-agile balancing components bring autonomy and flexibility to *ART* and teams since they can decide how they distribute funding to development activities without cost control centralization by the portfolio management, but at the same, it brings rigidity to enable coordination among ART and teams. Horizon planning is based on *Weighted Shortest Job First* (WSJF), a model designed to maximize financial output. Following this approach, business owners favor *epics and enablers*[4] that produce a fast value.

In a traditional approach, funding is distributed to each project or teams where portfolio managers centralize decision-making for updating budget in case of changes. This centralization has been accused of cost overruns, delays, blaming culture and teams' demotivation. For SAFe, funding is directly distributed to value streams where teams involved in the same value stream or ART can decide how funding is then distributed to the *epics, enablers*, and *user stories*. This decentralized distribution is undertaken during the *Innovation and planning iteration*[5].

SAFe also proposes lean and agile principles for *portfolio performance* (goal 4), including portfolio monitoring and reporting. Business drivers and *enterprise architects* use two specific artifacts: the *portfolio Kanban* and portfolio roadmap. The Kanban aims to illustrate *epics* throughout the portfolio process: 1) epic ideation, 2) epic review, 3) epic analysis, 4) portfolio backlog (including selected and prioritized *epics*), 5) *epics* implementation and 5) epic done. Kanban permits to visualize *epics* flow with the goal of enabling collaborative, transparent and accurate decision-making for the portfolio management. Furthermore, organization can use this artifact to match *portfolio Kanban* cadence to the organizational capacity, align expectations and minimize work in progress for teams and *ARTs*. On the other hand, the *portfolio roadmap* visualizes how *ARTs* and teams

216 • *Portfolio Management*

achieve the portfolio vision over different periods of time, usually *epics* are illustrated in function of the four-planning horizon explained before.

Overarching Mechanisms for Portfolio: Governance

SAFe proposes governance practices to sustain portfolio goals (strategic direction, value maximization, portfolio balancing and portfolio performance) with flexibility and rigidity (Almeida & Espinheira, 2021; Kowalczyk et al., 2022). Portfolio governance encompasses the set of procedures used by the organization to oversee portfolio management. Project governance defines practices and tools to bring the right information to the right people, at the right time, to enable them to make collaborative decisions. Strong governance also ensures the consistency and replicability of portfolio management activities. SAFe portfolio governance considers ceremonies to enable continuous review of the portfolio, guardrails to enable portfolio management and specific portfolio roles.

Portfolio governance considers collaboration as a key element to sustain portfolio vision and decision-making. Indeed, collaboration provides a need for the pursuit of common organizational goals, discussions across domains and recurring feedback opportunities. To enable collaboration, SAFe defines specific roles and responsibilities for actors involved in the portfolio management: business owners who are responsible for defining strategic themes and maintain value streams, business architects who are responsible for defining value streams and lean portfolio managers who have the highest authority within the organization for enabling portfolio goals and governing portfolio. The portfolio managers define *portfolio guardrails* and facilitate portfolio ceremonies to enable collaboration.

Portfolio guardrail aims to enable collaboration among business owners while they are balancing the portfolio following lean-agile budget means. Guardrails define how *epics* and *enablers* are categorized in four development horizons, how funding is assigned to value streams considering organizational capacity and how significant changes can be approved. *Guardrails* look to maintain business owners' engagement to the portfolio strategy by ensuring that teams use funding assigned to *ARTs* on the right *epics* and *enablers*.

For SAFe, portfolio management is not static, and portfolio goals are not executed on an annual basis. Indeed, agile portfolio management considers frequently reviewing whether the teams and/or *ART* deliverables are aligned with portfolio vision and if not, immediate changes are introduced in the *value streams*, ARTs and funding distribution. SAFe proposes three different ceremonies that enable continuously portfolio monitoring, opportunities evaluation, and accountability. First, a *strategic portfolio review* is a ceremony that takes place approximately every 3 months, usually in the middle of a *Program Increment*[6] (PI). During this meeting, strategic themes and their progress are inspected, and any adjustments considered can be applied. Business drivers review and modify, if necessary, the portfolio vision, strategic themes, investments by horizons, guardrails, metrics, and portfolio roadmap. Second, a *portfolio synchronization* ceremony aims to inspect current initiatives regarding the strategic theme. With a monthly cadence, business drivers review portfolio hypotheses to decide whether to cancel, pivot or persevere *value streams*. They can update the *portfolio Kanban* with the status of the *epics*, review blockers and impediments of the ARTs that may affect the achievement of strategic objectives, collect metrics' data and KPIs of each *value stream* and update the *portfolio roadmap*. Finally, a *participatory budgeting* ceremony takes place twice a year where members from different *value streams* propose, from their knowledge, how they would distribute the budgets. This dynamic helps those responsible for budget allocation to consider the expert opinions of all areas of the business lines, to increase their involvement and consensus in decision-making.

SCALING FRAMEWORK SAFe IN PRACTICE

The SAFe framework is prescriptive. It provides numerous guidelines and recommendations which were presented in the previous section. Although the number of documents on SAFe (or on APM under other agile frameworks) deployment in academic journals is still limited (Laanti & Kangas, 2015; Lindkvist et al., 2017; Paasivaara, 2017; Puthenpurackal Chakko et al., 2021; Rautiainen et al., 2011), we had access to the empirical data on a SAFe deployment case at a large South-African bank where the research focus was on strategic alignment of information technology initiatives in a scaled agile environment (Petit & Marnewick, 2021).

Overview of the Case

Petit and Marnewick (2021) published the results of a research investigating how initiatives are aligned with an organization's strategy in a scaled agile environment. The implementation of SAFe was studied in 2018 at the Bank[7], currently among the largest organizations in South Africa by market capitalization. It offers a wide range of banking and financial services in 20 countries in Africa. The focus of the research was Group IT which is a common function to the Bank and employs over 6 000 people. It is a centralized unit offering IT expertise and development capability to approximately seven units/divisions, each with its own internal objectives, structure and portfolio.

The company began the implementation of SAFe in 2015 to be closer to the business, to deliver products in weeks rather than years, to build more usable/simple software and to adopt new technologies faster. Nineteen interviews were conducted with various individuals within the Bank's Group IT division. The results of this research (Petit & Marnewick, 2021) serve, in this chapter, as a practical case to compare how a large organization has been trying to reach the goals of PPM and governance in a scaled agile environment in comparison to the SAFe framework described above.

Strategic Direction

The executive committee of the Bank developed a five-year vision supported by a strategy composed of eight themes. Most of the themes were expressed in one to five words, but would affect all units, one way or another. These themes had the advantage of being simple to remember and easy to relate to. Surprisingly, most of the participants to this research could easily refer back to them.

SAFe framework suggests collaboration from all the business owners to define strategic themes and get business units' engagement to the portfolio alignment. This was observed in most of the business units' cases, *Business Owners* recognized that the portfolio strategy was aligned with the organizational strategy. However, some business units did feel that although there was some alignment, it was still not yet 100% aligned. This could be attributed to still limited business unit's SAFe implementation.

Value Maximization

Although the SAFe framework suggests linking the strategic themes to the organizational *value streams*, the Bank acknowledged that the notion of *value streams* had not yet been deployed. The CIO considered that this process of value maximization was performed prior to the SAFe implementation when projects were selected, and resources allocated according to financial criteria to the different projects. In their traditional portfolios, the bank never went back to assess value and benefits.

This remained similar during their SAFe implementation. The CIO felt that it was not possible to control and command the development of the initiatives from the top. They were convinced that this "would not happen" at the Bank.

In summary, the SAFe framework suggests that the organizational vision should be implemented through various strategic themes and deployed and followed up using *value streams*. In the case of the Bank the translation of strategic themes into *value streams* was not observed. The implication was that no value was attached to a specific theme, making it difficult to balance the portfolio, as discussed in the following section. However, *epics* and *ART* were used extensively.

Portfolio Balance

One of the purposes of portfolio management is the optimal balance of all the respective components that forms part of the portfolio. The selection and continuous balancing of the various components contributes to the success of the portfolio. SAFe framework suggests balancing portfolio defining guardrail, following horizon plans and participatory funding. *Guardrails* were not used at the Bank but other practices were used.

At the highest hierarchical level of the Bank, budgets were allocated to the Business Units for their respective product development. This was the first form of balancing. A second form of balancing, introduced with SAFe, was *PI* planning exercises within each business unit. One of the main benefits of the PI planning exercise was the visibility of the backlog to be developed for a given *PI*. During the *PI planning*, each of the features displayed on the wall in the planning room was color coded (e.g., red, yellow, blue, green) in alignment with the strategic themes. It was therefore possible to visualize how the strategic themes would be balanced just by standing back and looking at the wall and the colors. One of the interviewees at the Bank described a meeting where a product owner rebalanced the portfolio by increasing some of the "yellow" strategic theme items at the expense of other colors to ensure a correct balance (from a strategic point of view).

Portfolio Performance

The Bank's board of directors has instituted a subcommittee to oversee the delivery of IT. The portfolio management office had to report to that subcommittee, for example on the outcomes of the IT investments. In the past, they reported on project performance. With the introduction of SAFe, they have now started to report on the strategic themes set by the Bank. They can only do this by collecting the results of the *PI* deliveries linked back to the strategic themes.

Although many of the interviewees at the Bank stated that they were reaping the benefits of scaling agile in the organization, at the time of the study, benefits tracking (and *value stream* management) was non-existent, and the Bank was trying to put measures in place to address this weakness. They were in the middle of a process of deploying a tool to associate the data for each initiative to the strategy. In comparison, when they were previously in a project-based paradigm it was extremely difficult to link project completion to the benefits.

Portfolio Governance

SAFe framework proposes guardrails and ceremonies to enable portfolio governance and collaboration. However, the Bank sustains mainly portfolio with *PI planning*. Although the description of the *PI Planning* process above might make it sound as if the strategy is cascaded down in the organization, this is not exactly how this was observed in practice. In practice, the specific business unit themes were indeed defined, based on the corporate theme. This could be considered top-down approach. However, IT initiatives identified by defining the expected outcomes to be delivered in the next quarter for each *PI Product Owners* were then linked back to the business unit strategy. The *PI* plan linked the work to be delivered and the objectives, which translated to value, which then translated to the business strategy map. There was continuous evaluation of the various components for inclusion or exclusion in a *PI* based on the strategic alignment of the various components.

Although the corporate strategies inform the business units' portfolio strategies through a top-down approach, many instances were presented where a bottom-up approach had been followed. As a *Business Manager* put it: "*We don't want the bosses saying what needs to be done, we want it to come more bottom up [...] So we don't have portfolio meetings*" (Petit & Marnewick, 2021, p. 14).

There was one *Business Owner* for each business unit. They were ultimately responsible for the resource allocation and prioritization of the initiatives for that *Business Unit.* They were supported by a number of *Product Owners* who had the key role of ensuring that there was a constant alignment with the strategic themes by simultaneously collaborating with the business executives and being active at every *pre-PI* and *PI* planning session.

CONCLUSIONS

This chapter aimed at a better understanding of agile portfolio management. This topic has started gaining some interest in the literature but remains to be further explored in practice. Traditional portfolio management and agile portfolio management differ in several important aspects, as agile practices bring specific challenges at the portfolio levels and a need for a different approach. Practitioners have proposed several frameworks to deploy agility throughout the entire organization. Three frameworks stand out as most mature for portfolio management (LeSS, DAD and SAFe), but the most implemented in organizations is SAFe. Most importantly, this framework enables strategic direction, value maximization, portfolio balance and performance by adopting specific practices to govern and manage the portfolio such as: artifacts, guardrails and ceremonies.

However, it seems that the SAFe framework is often not yet fully established in practice. Our case study illustrates how a large bank implemented some artifacts and ceremonies to accomplish portfolio goals and governance. However, many others are not used, such as *value streams*, benefits realization, participatory funding, among others. In that case, PI planning seems to be the most used practice to ensure portfolio balance, performance and governance. Furthermore, SAFe framework suggests enabling collaboration for all the portfolio aspects to ensure alignment and engagement, but there is no evidence in our case of collaborative practices at the highest level. It is important to consider that the studied organization has transformed their portfolio practices from traditional to agile. During the study, the organization was still in a transformation journey. Traditional practices seemed to limit how the organization was implementing portfolio agile artifacts, guardrails and ceremonies.

This chapter makes three key contributions: 1) proposing four goals and an overarching mechanism for portfolio management that can be applied in both approaches (traditional or agile). This analytical framework guided our comparison of both approaches and served to get a better understanding of portfolio management regardless of the development approach. As portfolio management is about the value of projects to the organization, focusing on goals rather than processes makes more sense, especially given that agile principles invite us to depart from processes, 2) providing a better understanding of a widespread scaled agile framework such as SAFe, both in theory (through the analysis of this framework guided by previously mentioned goals and mechanism) and in practice (through a concrete exploration of a SAFe implementation) and 3) illustrating challenges to transform portfolio practices from traditional to agile framework and limits for implementing the SAFe framework for portfolio management.

However, this chapter also has limits: 1) A previous field research has been mobilized for the investigation of SAFe in practice. As the agile at scale in that organization has continued to increase at an important rate in the last years, more recent empirical work would be valuable. 2) For the purpose of this chapter, a single framework has been investigated. Although it is the most widespread and we provided a quick overview of a few other frameworks, it remains a partial view, especially that these frameworks have been criticized for being rigid and top-down (Sweetman & Conboy, 2018) and that many organizations do not follow a unique framework (Stettina & Hörz, 2015) or even develop their own large-scale agile method (Edison et al., 2021). As such, we recommend future research to explore other contexts longitudinally to understand how these frameworks are adapted and evolve in practice. 3) This chapter illustrates how agile portfolio management can be implemented for an organization where all the projects followed agile methodologies. However, this is not totally reflected in organization contexts where hybrid methods are promoted to execute projects (Gemino et al., 2021).

Along with these suggestions, we propose two key lines of inquiry to enrich the discussion on agile portfolio management. 1) Consider product management as an alternative to portfolio management (and what it means for the project management field). 2) Understand how hybridity (the mix of traditional and agile practices) plays out, and with what effects.

This is especially relevant as keeping "old processes" brings challenges to a successful agile transformation (Dikert et al., 2016).

In summary, this chapter offers an overview of how agile approaches accomplish portfolio management goals and governance in comparison to traditional approaches. Delving into SAFe, the most widespread framework for APM, both in theory and practice, this chapter helps to drive forward a topic that will only gain in relevance through time.

NOTES

1 An Epic is "a container for a significant solution development initiative that captures the more substantial investments that occur within a portfolio" (Scaled Agile, 2020, s.n.).
2 The Minimum Value Product (MVP) is a method that aims to define product scope in function of the most expected function (the killer feature) by a target audience, and to offer a product as quickly as possible to release it as soon as possible.
3 An Agile Release Train is a group of people (typically between 50 and 125) working in a coordinated manner and aligned with the business objectives in a constant flow to enable value for a Value Stream. It means that an ART is a project team made up of several agile teams, operations teams and specialists and a coordination team at the program level called the Release Management Team.
4 An enabler is a set of activities required to make a specific outcome that doesn't bring value to customer, but that enables exploring a new user story (spike), developing an architectural or infrastructure requirement, or compliance.
5 A planning and innovation iteration aims to define and plan a program increment (PI).
6 A program increment (PI) is a box of time during which an ART delivers incremental value in the form of working and tested solutions. PIs usually last between 8 and 12 weeks. The most common pattern for an IP is four development iterations, followed by an Innovation and Planning (IP) iteration.
7 Fictitious name to preserve anonymity.

REFERENCES

Agile Alliance. (2001). *Manifesto for Agile Software Development.* http://agilemanifesto.org/

Ahmad, M. O., Lwakatare, L. E., Kuvaja, P., Oivo, M., & Markkula, J. (2017). An Empirical Study of Portfolio Management and Kanban in Agile and Lean Software Companies. *Journal of Software: Evolution and Process, 29*(6), 1–16.

Almeida, F., & Espinheira, E. (2021). Large-Scale Agile Frameworks : A Comparative Review. *Journal of Applied Sciences, Management and Engineering Technology, 2*(1), 16–29.

Alqudah, M., & Razali, R. (2016). A Review of Scaling Agile Methods in Large Software Development. *International Journal on Advanced Science, Engineering and Information Technology, 6*, 828.

Ambler, S. W. (2017). *An Executive's Guide to Disciplined Agile: Winning the Race to Business Agility.* Scotts Valley, CA: CreateSpace Independent Publishing Platform.

Ambler, S. W. (2020). *Introduction to Disciplined Agile Delivery* (2nd Edition). Scotts Valley, CA: CreateSpace Independent Publishing Platform.

Ambler, S. W., & Lines, M. (2016). *The Disciplined Agile Process Decision Framework*, Paper presented at the Software Quality the Future of Systems and Software Development - 8th International Conference, SWQD 2016, Vienna (Austria), 3–16.

Axelos. (2011). *Management of Portfolios.* London, UK: Stationery Office Books.

Benko, C., & McFarlan, F. W. (2003). *Connecting the Dots: Aligning Projects with Objectives in Unpredictable Times.* Boston, MA: Harvard Business School Press.

Beringer, C., Jonas, D., & Kock, A. (2013). Behavior of Internal Stakeholders in Project Portfolio Management and its Impact on Success. *International Journal of Project Management, 31*(6), 830–846.

Conforto, E. C., & Amaral, D. C. (2016). Agile Project Management and Stage-Gate Model—A Hybrid Framework for Technology-Based Companies. *Journal of Engineering and Technology Management, 40*, 1–14.

Cooke-Davies, T. J., Crawford, L. H., & Lechler, T. G. (2009). Project Management Systems: Moving Project Management from an Operational to a Strategic Discipline. *Project Management Journal, 40*(1), 110–123.

Cooper, R. G. (2008). Perspective: The Stage-Gate Idea-to-Launch Process—Update, What's New, and NexGen Systems. *The Journal of Product Innovation Management, 25*(3), 213–232.

Cooper, R. G., Edgett, S. J., & Kleinschmidt, E. J. (1997a). Portfolio Management in New Product Development: Lessons from the Leaders—I. *Research Technology Management, 40*(5), 16–28.

Cooper, R. G., Edgett, S. J., & Kleinschmidt, E. J. (1997b). Portfolio Management in New Product Development: Lessons from the Leaders—II. *Research Technology Management, 40*(6), 43–52.

Cooper, R. G., Edgett, S. J., & Kleinschmidt, E. J. (1998). Best Practices for Managing R&D Portfolios. *Research Technology Management, 41*(4), 20–33.

Cooper, R. G., Edgett, S. J., & Kleinschmidt, E. J. (2001). *Portfolio Management for New Products* (2nd Edition). Cambridge, MA: Perseus.

Cooper, R. G., Edgett, S. J., & Kleinschmidt, E. J. (2002a). Optimizing the Stage-Gate Process: What Best-Practice Companies Do - Part 1. *Research Technology Management, 45*(5), 21–27.

Cooper, R. G., Edgett, S. J., & Kleinschmidt, E. J. (2002b). Optimizing the Stage-Gate Process: What Best-practice Companies Do - Part 2. *Research Technology Management, 45*(6), 43–49.

De Piante Henriksen, A., & Jensen Traynor, A. (1999). A Practical R&D Project-Selection Scoring Tool. *IEEE Transactions on Engineering Management, 46*(2), 158–170.

Digital.ai. (2021). *15th State of Agile Report.*

Dikert, K., Paasivaara, M., & Lassenius, C. (2016). Challenges and Success Factors for Large-Scale Agile Transformations : A Systematic Literature Review. *Journal of Systems and Software, 119*, 87–108.

Dingsøyr, T., Fægri, T. E., & Itkonen, J. (2014). What Is Large in Large-Scale? A Taxonomy of Scale for Agile Software Development. In *Product-Focused Software Process*

Improvement: Vol. 8892 of the series Lecture Notes in Computer Science (pp. 273–276). Helsinki, Finland: Springer.

Dinsmore, P. C., & Cooke-Davies, T. J. (2006). *Right Projects Done Right! : From Business Strategy to Successful Project Implementation* (24119021). San Francisco, CA: Jossey-Bass.

Dye, L. D. (2002). *Using Scenario Planning as an Aid in Project Portfolio Management.* San Antonio (TX): PMI Global Congress North America.

Edison, H., Wang, X., & Conboy, K. (2021). Comparing Methods for Large-Scale Agile Software Development: A Systematic Literature Review. *IEEE Transactions on Software Engineering, 48*(8).

Fitzgerald, B., & Stol, K.-J. (2017). Continuous Software Engineering: A Roadmap and Agenda. *Journal of Systems and Software, 123,* 176–189.

Garfein, S. J. (2005, September 10). *Strategic Portfolio Management: A Smart, Realistic and Relatively Fast Way to Gain Sustainable Competitive Advantage.* Toronto: PMI Global Congress North America.

Gemino, A., Horner Reich, B., & Serrador, P. M. (2021). Agile, Traditional, and Hybrid Approaches to Project Success: Is Hybrid a Poor Second Choice? *Project Management Journal, 52*(2), 161–175.

Grundler, A., & Westner, M. (2019). *Scaling Agile Frameworks vs. Traditional Project Portfolio Management: Comparison and Analysis.* Proceedings of the International Conferences on Internet Technologies & Society (ITS 2019), Hongkong, 51–62.

Gustafsson, J., & Salo, A. (2005). Contingent Portfolio Programming for the Management of Risky Projects. *Operations Research, 53*(6), 946–956.

Hermano, V., & Martin-Cruz, N. (2016). The Role of Top Management Involvement in Firms Performing Projects : A Dynamic Capabilities Approach. *Journal of Business Research, 69,* 3447–3458.

Hobbs, B., & Petit, Y. (2017a). *Agile Approaches on Large Projects in Large Organizations.* Newton Square, PA: Project Management Institute.

Hobbs, B., & Petit, Y. (2017b). Agile Methods on Large Projects in Large Organizations. *Project Management Journal, 48*(3), 3–19.

Hodgkins & Hohmann. (2007). *Agile Program Management: Lessons Learned from the VeriSign Managed Security Services Team. Agile 2007 (AGILE 2007),* 194–199.

ISO. (2015). Project, Programme and Portfolio Management—Guidance on Portfolio Management. *Standard,* 19.

Kalliney, M. (2009). *Transitioning from Agile Development to Enterprise Product Management Agility. 2009 Agile Conference,* 209–213.

Kaufmann, C., Kock, A., & Gemünden, H. G. (2020). Emerging Strategy Recognition in Agile Portfolios. *SI: Actors, Practices and Strategy Connections in Multi-Project Management, 38*(7), 429–440.

Killen, C. P., Jugdev, K., Drouin, N., & Petit, Y. (2012). Advancing Project and Portfolio Management Research : Applying Strategic Management Theories. *International Journal of Project Management, 30*(5), 525–538.

Killen, C. P., Jugdev, K., Drouin, N., & Petit, Y. (2013). Translational Approaches: Applying Strategic Management Theories to OPM Research. In N. Drouin, R. Müller, & S. Sankaran (Éds.), *Novel Approaches to Organizational Project Management Research: Translational and Transformational* (001227634; p. 348–378). Koege, Denmark: Copenhagen Business School Press.

Kowalczyk, M., Marcinkowski, B., & Przybyłek, A. (2022). Scaled Agile Framework. Dealing with Software Process-Related Challenges of a Financial Group with the Action Research Approach. *Journal of Software: Evolution and Process, 34*(6), e2455.

Laanti, M., & Kangas, M. (2015). *Is Agile Portfolio Management Following the Principles of Large-Scale Agile? Case Study in Finnish Broadcasting Company Yle, 2015 Agile Conference* (pp. 92–96). IEEE.

Lanka, M. (2007). *Strategically Aligning Your Project Portfolio*. Atlanta (GA): PMI Global Congress.

Lan-Ying, D., & Yong-Dong, S. (2007). Implement Business Strategy via Project Portfolio Management : A Model and Case Study. *Journal of American Academy of Business, Cambridge, 11*(2), 239–244.

Larman, C., & Vodde, B. (2014). *Large-Scale Scrum: More with Less*. Boston, MA: Addison Wesley Professional.

Leffingwell, D. (2017). *SAFe® 4.0 Reference Guid : Scaled Agile Framework® for Lean Software and Systems Engineering*. Boston, MA: Addison-Wesley Professional.

Levin, G., & Wyzalek, J. (2014). *Portfolio Management: A Strategic Approach*. Boca Raton, FL:Auerbach Publications.

Lindkvist, L., Bengtsson, M., Svensson, D.-M., & Wahlstedt, L. (2017). Replacing Old Routines: How Ericsson Software Developers and Managers Learned to Become Agile. *Industrial and Corporate Change, 26*(4), 571–591.

Martinsuo, M., Gemünden, H. G., & Huemann, M. (2012). Toward Strategic Value from Projects. *International Journal of Project Management, 30*, 637–638.

Martinsuo, M., & Lehtonen, P. (2007). Role of Single-Project Management in Achieving Portfolio Management Efficiency. *International Journal of Project Management, 25*(1), 56–65.

McDonough, E. F., & Spital, F. C. (2003). Managing Project Portfolios. *Research Technology Management, 46*(3), 40–46.

Milosevic, D. Z., & Srivannaboon, S. (2006). A Theoretical Framework for Aligning Project Management with Business Strategy. *Project Management Journal, 37*(3), 98–110.

Morris, P. W. G., & Jamieson, A. (2005). Moving from Corporate Strategy to Project Strategy. *Project Management Journal, 36*(4), 5–18.

Müller, R., Martinsuo, M., & Blomquist, T. (2008). Project Portfolio Control and Portfolio Management Performance in Different Contexts. *Project Management Journal, 39*(3), 28–42.

Narayan, S. (2015). *Agile IT Organization Design: For Digital Transformation and Continuous Delivery*. Boston, MA: Addison-Wesley Professional.

Nielsen, K. (2021). *Agile Portfolio Management : A Guide to the Methodology and Its Successful Implementation "Knowledge That Sets You Apart"*. New York, NY: Productivity Press.

Paasivaara, M. (2017). *Adopting SAFe to Scale Agile in a Globally Distributed Organization, 2017 IEEE 12th International Conference on Global Software Engineering (ICGSE)* (pp. 36–40), Buenos Aires, Argentina, 2017.

Pennypacker, J. S., & Dye, L. D. (2002). *Managing Multiple Projects: Planning, Scheduling, and Allocating Resources for Competitive Advantage*. New York, NY: Marcel Dekker, Inc.

Petit, Y., & Hobbs, B. (2010). Project Portfolios in Dynamic Environments: Sources of Uncertainty and Sensing Mechanisms. *Project Management Journal, 41*(4), 46–58.

Petit, Y., & Hobbs, B. (2012). Project Portfolios in Dynamic Environments: Organizing for Uncertainty. *International Journal of Project Management, 30*(5), 539–553.

Petit, Y., & Marnewick, C. (2021). Strategic Alignment of Information Technology Initiatives in a Scaled Agile Environment. *The Journal of Modern Project Management 8* (3), 6–29.

Project Management Institute. (2016). *Governance of Portfolios, Programs, and Projects—A Practice Guide.* Newton Square, PA: Project Management Institute.

Project Management Institute. (2017). *The Standard for Portfolio Management* (4th Edition). Newton Square, PA: Project Management Institute.

Puthenpurackal Chakko, J., Huygh, T., & De Haes, S. (2021). Achieving Agility in IT Project Portfolios – A Systematic Literature Review. In A. Przybyłek, J. Miler, A. Poth, & A. Riel (Éds.), *Lean and Agile Software Development* (pp. 71–90). Springer International Publishing.

Rad, P. F., & Levin, G. (2005). A Formalized Model for Managing a Portfolio of Internal Projects. *AACE International Transactions*, PM41-45.

Rad, P. F., & Levin, G. (2007). *Project Portfolio Management Tools and Techniques.* New York, NY: International Institute for Learning.

Rautiainen, K., von Schantz, J., & Vahaniitty, J. (2011). *Supporting Scaling Agile with Portfolio Management: Case Paf.com*, 1–10.

Rigby, D. K., Sutherland, J., & Noble, A. (2018). Agile at Scale. *Harvard Business Review, 96*(3), 88–96.

Royer, I. (2003). Why Bad Projects Are so Hard to Kill. *Harvard Business Review, 81*(2), 48–57.

Scaled Agile. (2020). SAFe5 for Lean Enterprises. *SAFe.* https://www.scaledagileframework.com/

Scheerer, A., Hildenbrand, T., & Kude, T. (2014, January). Coordination in large-scale agile software development: A multiteam systems perspective. In *2014 47th Hawaii international conference on system sciences* (pp. 4780–4788). IEEE.

Shenhar, A. J., Milosevic, D., Dvir, D., & Tamhain, H. (2007). *Linking Project Management to Business Strategy.* Newton Square, PA: Project Management Institute.

Stettina, C. J., & Hörz, J. (2015). Agile Portfolio Management: An Empirical Perspective on the Practice in Use. *International Journal of Project Management, 33*(1), 140–152.

Sweetman, R., & Conboy, K. (2018). Portfolios of Agile Projects: A Complex Adaptive Systems' Agent Perspective. *Project Management Journal, 49*(6), 18–38.

Thomas, J. C., & Baker, S. W. (2008, August). *Establishing an Agile Portfolio to Align IT Investments with Business Needs*, In *Agile 2008 Conference*, IEEE. (pp. 252–258).

Uludag, O., Kleehaus, M., Xu, X., & Matthes, F. (2017). *Investigating the Role of Architects in Scaling Agile Frameworks. 2017 IEEE 21st International Enterprise Distributed Object Computing Conference (EDOC)*, 123–132.

Vaidya, A. (2014, October 20). *Does DAD Know Best, Is it Better to Do LeSS or Just Be SAFe? Adapting Scaling Agile Practices into the Enterprise. Pacific Northwest Software Quality Conference (PNSQC)*, Portland (OR).

12

Dynamic Capability through Project Portfolio Management

Catherine P. Killen

INTRODUCTION

During the past two decades, project portfolio management (PPM) has become established as a discipline and organizations have increasingly turned to PPM to help them manage their portfolios of projects and improve their competitive position (Levine, 2005; Kester et al., 2011; Killen & Drouin, 2017). The primary goals of PPM adoption are to effectively implement the organizational strategy through the portfolio of projects and to enhance the long-term value of the portfolio. PPM assists with the management of resources across the portfolio to avoid a "resource crunch," where the organization attempts too many projects (Cooper & Edgett, 2003). PPM methods also provide the holistic oversight required to ensure balance in the portfolio. The link between higher success levels and the use of formal and mature PPM approaches (Cooper et al., 2001; Killen et al., 2008a) has prompted organizations to focus on the establishment and development of PPM.

PPM is often presented as a series of processes and procedures that organizations tailor to suit their environment. The common refrain has been that once tailored appropriately, the PPM process will assist an organization in achieving competitive advantage by implementing strategy, balancing the portfolio, maximizing value, and ensuring resource adequacy for projects. However, recent research highlights many other aspects of PPM that paint a picture of increased complexity and dynamism and offers insight into additional ways that PPM can create value for an organization (Killen & Hunt, 2010; Petit, 2012). PPM is now seen as more than a process; PPM is an organizational capability that also includes the organizational structure, the people, and the culture. These elements must work together for effective PPM and top management support is a crucial factor in PPM capability success. Recent studies also indicate that PPM has a significant role to play in helping organizations achieve advantages in dynamic environments, and that the PPM capability itself needs to evolve and adjust to enhance organizational agility and contribute to sustainable competitive advantage (Killen & Hunt, 2010, Sicotte et al., 2014, Killen & Drouin, 2017, Saeed et al., 2021).

This chapter first introduces PPM concepts and outlines typical processes before discussing the additional challenges for PPM in dynamic environments. To guide practitioners, several examples are presented to illustrate aspects of PPM that help organizations respond to change and improve organizational outcomes.

PPM CONCEPTS

As many organizations shift to "management by projects," projects are often the main vehicle for delivering organizational strategy. Definitions of PPM have evolved as the discipline became established. A widely accepted definition of PPM developed by Cooper et al. (2001, p. 3) is that:

> Portfolio management ... is a dynamic decision process wherein the list of ... projects is constantly revised. In this process, new projects are evaluated, selected, and prioritized. Existing projects may be accelerated, killed, or deprioritized and resources are allocated and reallocated.

McDonough and Spital (2003, p. 40) point out that PPM is more than project portfolio selection as it also involves the

> day to day management of the portfolio including the policies, practices, procedures, tools and actions that managers take to manage resources, make allocation decisions and ensure that the portfolio is balanced in such a way to ensure successful portfolio-wide new product performance.

Levine (2005, p. 22) offers a broad definition of PPM: "Project portfolio management is the management of the project portfolio so as to maximize the contribution of projects to the overall welfare and success of the enterprise." An organization's capability to manage the project portfolio encompasses much more than the processes and methods identified for PPM, it also requires the people and a culture that support information transparency and portfolio-level perspectives, and it requires organizational structures that provide appropriate levels of visibility and responsibility to support the PPM capability (Killen & Hunt, 2010). In addition to delivering strategy, dynamic PPM capabilities contribute to the evolution of strategy through a two-way interaction between the projects and the strategy. Project portfolio perspectives provide the impetus to update strategies to reflect dynamic environments and to recognize the emergence of bottom-up strategies (Kopmann et al., 2017; Kaufmann et al., 2020).

Although PPM needs to be tailored for each organization, there are many common elements and approaches to PPM. In its simplest form, PPM facilitates decisions across the entire portfolio of projects by collecting information from all the projects (both existing and proposed); collating and organizing the information; presenting information to a carefully selected decision-making team for portfolio-level review; and providing a structure for communicating and implementing decisions. These four steps are explained with extensions for dynamic environments in the section *Outline of a Dynamic PPM Approach.*

Figure 12.1 illustrates a range of common methods and tools for organizing and presenting portfolio data in a visual format, including (clockwise from top left) risk versus reward portfolio map, scoring model, dashboard display, stoplight report, and pie chart. Portfolio mapping is a common method to provide a central view of all projects in the portfolio to support decision-making. Portfolio maps plot projects on two axes and can assist with the selection of a balanced portfolio of projects. Commonly used portfolio maps balance aspects such as risk versus return and can also display other information through the size, color, patterns, or notes associated with the symbol for each project. Scoring models use weightings and ratings to compare projects based on multiple criteria. Many software applications for PPM offer "dashboard" displays that show the status of projects on dials and graphs. Stoplight reporting uses red and amber to highlight trouble areas and green to show the "all clear." While most visualizations of project portfolio data consider each project as an

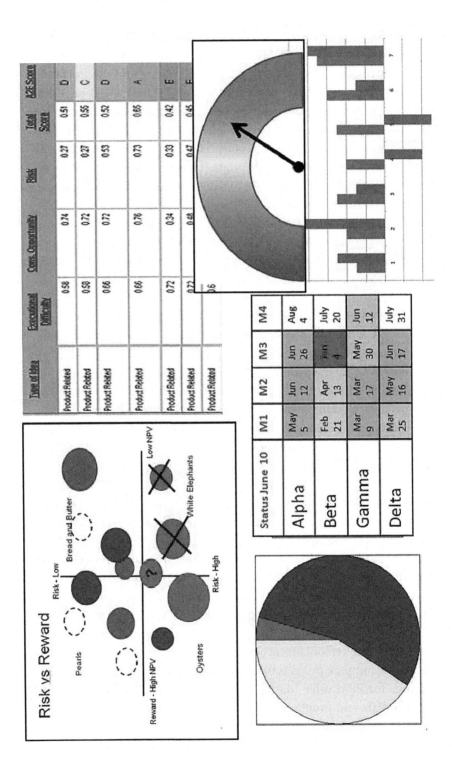

FIGURE 12.1
Typical methods for organizing and presenting PPM data.

independent entity, dependencies between projects can be highlighted through a dependency matrix or network mapping formats (Killen, 2022). Customizing PPM to best support decision-making in a specific environment usually involves selecting from common methods and then designing the process to include relevant parameters for each method.

Developing an effective dynamic PPM capability takes time; antecedent capabilities must be established before the PPM process can be established (Eisenhardt & Martin, 2000; Cooper et al., 2001). For example, establishing a foundational capability such as a gated project management process is an antecedent to the development of an effective PPM capability, and data gathering capabilities must be developed before the capability to evaluate and adjust the portfolio mix can be established (Martinsuo & Lehtonen, 2007).

As shown in Figure 12.2, PPM capabilities generally include a gated project management process integrated with a portfolio-level review process at one or more of the gates or decision points. The figure reflects the fact that many organizations develop more than one version of project management process to cater for different project types. The main differences between the versions are in the number of stages and gates and in the types of criteria used to evaluate projects at the gates. Also illustrated are the three main dimensions of a PPM capability: "process" dimensions, "structure" dimensions, and "people and culture" dimensions.

Figure 12.2 also depicts the post implementation review (PIR) as part of the process. The PIR is an important stage of the process because the "lessons learned" from each project enables the review, evaluation, and improvement of the project management and PPM processes. However, research indicates that this is a weak area in many organizations; it is common for managers to recognize the importance of PIR, but many find it difficult to allocate resources or gain support for such tasks (Killen et al., 2022).

PPM capabilities can improve organizational flexibility and performance by providing a holistic and responsive decision-making environment in dynamic environments. The role of the project portfolio manager is becoming formalized as organizations aim to gain the best results from PPM (Jonas, 2010). In addition to the challenge of multi-project management, organizations must address the challenges of an increasingly competitive, globalized, and deregulated environment characterized by shortening life cycles and dynamic markets. An organizational dynamic capability, the ability to adapt and respond to change, is essential in such rapidly changing environments (Killen & Hunt, 2010).

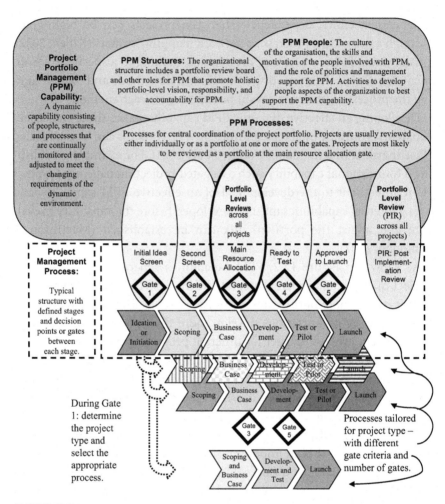

FIGURE 12.2
Three dimensions of PPM integrated with tailored gated project management processes.

A relatively recent concept of "agile" PPM (Stettina & Hörz, 2015; Hansen & Svejvig, 2018, Kaufmann et al., 2020, Ang et al., 2022), reflects the interest in PPM acting as a dynamic capability; however, the use of the term agile in association with PPM is contentious due to potential confusion with agile project management approaches. Agile project management approaches that offer an incremental and responsive approach to the management of projects are now common in an increasing range of environments; however, such approaches do not operate at the strategic project portfolio level. From the portfolio perspective, PPM can be "agile" and act as a dynamic capability by allowing an organization to identify changes in the environment and to evaluate, analyze, and adjust the portfolio to respond to changes in the environment. To observe changes in the environment, PPM requires a "sensing" capability that involves scanning the environment and regularly re-visiting assumptions (Teece, 2007). The PPM capability is responsible for configuring the organization's efforts by building and allocating resources. A PPM capability acts as a dynamic capability when it provides an organization with competitive advantages by enabling such reconfiguration in a timely fashion.

DYNAMIC CAPABILITIES AND COMPETITIVE ADVANTAGE THROUGH PPM

Dynamic capabilities are a special type of capability that enables an organization to respond to changes in the environment. Frameworks to identify and understand dynamic capabilities have emerged from research on strategy and competitive advantage. One of the goals of strategy research is to determine why some organizations are more successful than others and to understand the mechanisms that help organizations achieve a competitive advantage. PPM has been identified as one of these mechanisms (Killen et al., 2007; Killen & Hunt, 2010). Competitive advantage is the ability of an organization to create more value than its rivals and, therefore, achieve superior return on investment (Barney & Hesterly, 2012). One of the streams of strategy research is the resource-based view, which proposes that the differences in the levels and types of resources between competing organizations can be used to explain differences in organizational success rates. An extension or offshoot of the resource-based view is the identification of a special class of organizational capabilities that enable organizations to effectively respond to changes in the dynamic environments in which they compete (Teece et al., 1997). "Dynamic capabilities" do this by providing a capacity for "an organization to purposefully create, extend, or modify its resource base" (Helfat et al., 2007, p. 4).

The PPM capability of an organization is one of the internal organizational capabilities or resources that an organization uses to gain competitive advantage. In a dynamic environment, a PPM capability that acts as a dynamic capability can enable an organization to be agile and respond to changes in the environment. Improved innovation capabilities are also associated with dynamic PPM capabilities (Saeed et al., 2021). Although dynamic capabilities are a type of resource-based capability, they do not have the ability to create value independently. Dynamic capacities add value by working with the existing resource base (Eisenhardt & Martin, 2000) and can therefore be considered as "enabling resources" (Smith et al., 1996). It is also important that supporting capabilities are established before a dynamic capability can be effective (Eisenhardt & Martin, 2000). Therefore, a dynamic capability such as PPM must be accompanied by underlying resources and capabilities such as the project management capability to provide long-term competitive advantage in dynamic environments. Dynamic capabilities play a key role in allocating resources as well as in identifying the desired development and direction of resources and capabilities in line with strategy (Wang & Ahmed, 2007). As a dynamic capability, PPM can improve an organization's ability to "integrate, build, and reconfigure internal and external competencies to address rapidly changing environments" (Teece et al., 1997, p. 516) and through these mechanisms improve the competitive advantage in dynamic environments.

PPM IN DYNAMIC ENVIRONMENTS

Learning and change are an important part of the ability of PPM to provide an advantage in dynamic environments (Killen et al., 2008b). Figure 12.3 illustrates the effect of learning and change on the evolution of the PPM capability to meet the requirements of a dynamic environment. With learning and

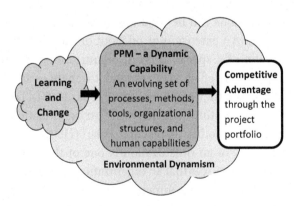

FIGURE 12.3
Learning and change: competitive advantage through the evolution of PPM in dynamic environments.

change, PPM can be a dynamic capability and enhance competitive advantage. Organizational learning is embedded in PPM capabilities through mechanisms for tacit and explicit learning. For example, tacit learning – the type of learning that is difficult to document or codify and is best transferred through experience or observation – is achieved through the interaction of experienced managers in PPM meetings and through the ability of PPM to act as a focal point for decision experiences to be shared and for learning to accumulate. On the other hand, explicit learning – the type of learning that can be codified and documented – is incorporated in PPM through aspects such as standard templates, databases, and defined and documented methods and routines. Both types of learning inform the evolution of the PPM capability and ensure that it remains up to date and relevant in a changing environment. Through this learning, the PPM process is able to deliver competitive advantage in dynamic environments.

OUTLINE OF A DYNAMIC PPM APPROACH

A typical portfolio-level review process is outlined in Figure 12.4 and includes four steps: single project data collection, portfolio data development, team decision-making, and implement decisions. In the following discussion, the general aspects of the four steps are detailed first followed by *specific aspects of PPM for dynamic environments in italics.*

Single Project Data Collection

Data is collected for new project proposals and on existing project status to inform decision-making (Kester et al., 2011). The data is generally collected from all relevant projects in a standard form that defines the types of data required to facilitate evaluation. Project data may be obtained from a computer system or through templates or proposal documents. Templates often include a one-page executive summary that highlights the main criteria for the decision-makers to consider; for example, risk, reward, investment, skills and resources required, benefits, and aims.

Dynamic environments may require more frequent refreshing of project data. The relevant types of data must be kept up to date and the templates for data collection may change periodically in response to capability reviews. In addition, beyond simply collecting project data, a dynamic PPM

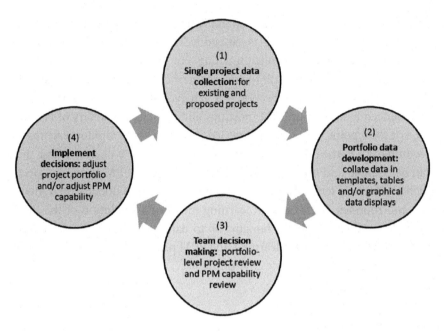

FIGURE 12.4

Outline of a dynamic PPM approach including evolution of processes through capability reviews.

capability may promote or encourage project ideas that support the organizational strategy. Idea management portals and collaborative tools can be used to assist in the idea and project proposal development process.

Portfolio Data Development

Drawing on the information for all projects in the portfolio, the data is collated or "rolled up" to provide portfolio-level summaries. The data is then arranged to assist decision-makers with the comparison and evaluation of portfolio data.

Research indicates that "best practice" organizations create graphical and visual information displays such as portfolio maps to facilitate group decision-making (Cooper et al., 2001; Mikkola, 2001; Killen et al., 2008a; de Oliveira Lacerda et al., 2011). Figure 12.1 illustrates some common portfolio-level data displays, including portfolio maps, which are developed in this stage of the process. Portfolio maps display projects and the strategic options they represent on two axes augmented with additional data to provide a visual representation that incorporates information such as strategic alignment, risk, return, and competitive advantage. Visuals boost the ability of decision-makers to understand and retain information by reducing the required cognitive load (Van der Hoorn, 2021). Due to the multiple types of data represented, these types of visual display are often called two-and-a-half dimensional (2½-D) displays (Warglien, 2010).

Such displays and all portfolio-level summaries must be kept up to date in dynamic environments. Some tools and techniques may be better suited to dynamic environments, and new tools and techniques are regularly developed and tested to meet current challenges. For example, network mapping approaches may help identify flow-on effects among interdependent projects arising from changes in the portfolio (Killen and Kjaer, 2012; Killen, 2017).

Team Decision-Making

In many organizations, a portfolio review board meets periodically to discuss the options available and to make project decisions in the context of the entire portfolio of projects (including ongoing projects as well as new proposals). A portfolio review board generally consists of five to ten experienced executives or managers that represent diverse organizational perspectives and responsibilities. There are many approaches to the timing of portfolio-level reviews; for example, some organizations create an annual portfolio plan, whereas others meet to refine the portfolio every week or so. The timing depends on the organization's environment and is influenced by aspects such as complexity, dynamism in the market, levels of technological change, and project duration. Meetings often employ graphical data representations to inform group discussions and negotiations (Mikkola, 2001; Killen et al., 2008a, Killen et al., 2020). Decisions are made with the entire portfolio in mind and the decision-making will consider resourcing, strategic alignment, balancing risk and other aspects. Typical decisions on new project proposals at a portfolio review meeting range from approval, hold for a later date, and rejection to requests for more information. The decisions relate to new projects as well as existing projects through mid-stream reviews (Rad & Levin, 2008); for example, ongoing projects can be canceled, delayed, accelerated, or left unchanged.

In dynamic environments, enhanced "sensing" capabilities need to be incorporated to detect changes in the environment; the time between decision meetings may need to be shorter; and special mechanisms may be required to enable agile response to unanticipated changes in the environment.

In addition, the regular review of the PPM capability is particularly important. The portfolio review board or other executives who make decisions about the projects in the portfolio must also review the processes used and their outcomes. The reviews track the results of the process and reflect on how the process worked. If necessary, recommendations for adjustments to the process are made. In a dynamic environment, regular reviews and adjustments will ensure that portfolio processes and outcomes reflect the evolving strategy.

Implement Decisions

The outcomes of the portfolio review decision meetings are then implemented. For example, new projects may be initiated; some existing projects may be canceled and resources reallocated; or other existing projects may be accelerated to beat the competition. These changes flow from the decisions made by the portfolio review board and continually adjust the portfolio of projects.

In dynamic environments, the adjustments to the portfolio may be more frequent. When the suggestions arising from the reviews of the PPM capability are implemented, the cycle continues with evaluation and adjustment of the capability as required. In dynamic environments, these reviews are built into the PPM process and drive the continual evolution of the processes for managing the portfolio.

EXAMPLES OF PPM IN DYNAMIC ENVIRONMENTS

What does a dynamic PPM capability look like in practice? The following examples are taken from a study of PPM approaches used by successful innovators. The examples were selected to illustrate practical examples of aspects of PPM capability that can improve organizational agility – an organization's ability to adjust to changes in the environment.

"Sensing" Changes in the Environment

"Sensing" changes in the environment is necessary for an organization to start the process of evaluating and adapting to those changes (Teece, 2007).

A medical devices company recognized the importance of keeping abreast of developments in medical treatments that could potentially influence their product development direction. The medical specialists employed by the organization played a significant role in "sensing" the environment; however, their time was limited and their expertise was focused in specific areas. Recognizing the importance of "sensing" the environment, the organization developed several strategies to keep track of trends and developments in the field. One of these strategies was the development of a medical review board, which consisted of external advisers and specialists from a range of related professions. This initiative greatly extended the available expertise and provided a diversity of perspectives.

Similarly, an approach employed by a telecommunications company was to encourage and facilitate employee involvement in specialist communities through conferences and professional associations. Through these contacts and conference presentations, the employees were better able to contribute to the organization's ability to "sense" the environment.

Reallocating Resources

An important aspect of PPM in any environment, especially dynamic environments, is the ability to stop poor projects and reallocate the resources to other projects. Organizations must be able to ensure that their project portfolio represents the best overall mix at the current time. Often a project that had strong support initially can become less desirable as the environment changes. The emergence of a new technology or competitive product, changes in demographics or foreign exchange rates, or changes in commodity or property prices can radically alter a project's prospects for success. However, many organizations find it difficult to cancel a poorly performing project and often the people involved resist changes to the project. One manufacturing organization felt that a culture that supported information and decision transparency and communication was key. They implemented steps to ensure that the criteria, data, and methods for evaluation were openly shared and discussed. In addition, all levels of management visibly supported and participated in the PPM processes. Through these measures, the organization gained strong buy-in and support for the process. With such support, the organization felt that decisions to cancel a project and reallocate resources were understood and supported, which made it easier to cancel projects when necessary.

242 • Portfolio Management

Ensuring Ambidexterity

In many industries, it is important that an organization successfully
"exploits" and "explores" at the same time – this is sometimes called
"organizational ambidexterity" (Tushman & O'Reilly, 1996; Tushman
et al., 2002). Exploitation projects are generally short-term, incremental,
or low-risk undertakings that are relied on for day-to-day improvements
in existing offerings or operations. In contrast, exploration projects are
long-term, high-risk, radical, or breakthrough initiatives that aim to create innovative new capabilities and offerings to bring the organization to
the next level. Collating data across the portfolio of projects through PPM
can provide an organization with the ability to determine the current balance of project types. This is often done using graphical data displays such
as portfolio maps or pie charts.

If an imbalance is found, PPM processes can help redress the balance
(Petro, 2017). For example, a digital services organization introduced
targeted idea generation activities to increase the number of radical ideas
when it realized that its portfolio was skewed toward "exploitation" over
"exploration." This type of skewing is common and is known as the
"success trap" because accumulated decision-making experiences can
reinforce support for short-term "exploitation" projects at the expense of
the long-term "exploration" projects that organizations believe are
essential for long-term success (March, 1991). As one manager in a
financial services organization commented during an interview: "Short-
versus long-term is most difficult to balance, especially with pressure to
turn around in a shorter term. Longer term no one gives you any credit for
and it is harder to get justification."

To address this problem an industrial machinery manufacture allocated
a set percentage of its budget for each type of project to ensure the
appropriate balance. Another approach is to develop a separate tailored
process with appropriate evaluation criteria to be used with long-term
explorative projects, as illustrated in Figure 12.2. This approach ensures
that innovative ideas and projects are not disadvantaged by having to meet
rigid criteria that are not appropriate for "exploration" projects.

Adjusting the Portfolio Review Board

The membership of the portfolio review board is an important part of
a PPM capability. In a dynamic environment, the profile of the portfolio review board members may need to be adjusted as the environment

changes. For example, one successful manufacturer traditionally had a strong engineering and technical influence on the review board. This served the organization well during its initial stages of developing a best-in-class technology and enabled it to extend its market internationally. However, as the international competitive environment evolved, the portfolio review decisions failed to incorporate marketing and customer-related input, which resulted in several technologically driven projects failing to find a market. On review of the situation, the organization decided to radically change the membership of the portfolio review board to include marketing experience across the main regions. This change allowed the portfolio to better reflect marketing requirements in the regions.

Reviewing and Developing PPM Methods and Tools

Dynamic project environments are often characterized by complexity; interdependency between projects; and constraints in the availability of skills and resources. In such environments, PPM is a complex multi-dimensional challenge and the PPM capability must evolve to stay relevant. The challenge is amplified by the presence of interdependencies as PPM is more than an extension or scaled-up version of project management, which makes the inter-project effects more complex and difficult to predict (Aritua et al., 2009). The management of interdependencies is an area of weakness for PPM (Elonen & Artto, 2003) and is one of many areas where new tools are tested. Practitioners and researchers continually refine existing methods and tools as well as develop and test new ones. For example, network mapping methods to manage project interdependencies have been tested and shown to support decision-making (Killen 2017; Killen & Kjaer, 2012).

CONCLUSION

In dynamic environments, PPM can act as a dynamic capability that enables organizations to respond to change as it manages and balances the portfolio holistically; aligns projects with strategy; and ensures adequate resourcing for projects to maximize the benefits from project investments.

A PPM capability requires more than tools and methods for evaluating and making decisions on project portfolio data, it also requires appropriate organizational structures, a supportive culture, and top management support. One of the major challenges facing organizations is implementing a PPM capability that is flexible and responsive to changes in the environment. Although there are many common elements identified in PPM processes, there is evidence that each organization must tailor its PPM process to suit the individual environment and the PPM capability must be able to adapt and adjust to reflect changes in the environment.

A dynamic PPM capability can help project-based organizations develop a competitive advantage by responding to changes in the environment. Learning and change have been shown to be a vital component of a dynamic PPM capability, and several examples of PPM capability aspects that enhance agility were outlined. Practitioners can draw on these examples to stimulate ideas and gain organizational advantages through the development of a dynamic capability using PPM.

REFERENCES

Ang, K. C. S., Hansen, L. K., & Svejvig, P. (2022). Value-orientated decision-making in agile project portfolios. In R. Ding, R. Wagner, & C.-N. Bodea (Eds.), *Research on project, programme and portfolio management: Projects as an arena for self-organizing* (pp. 49–64). Cham (Switzerland): Springer International Publishing.

Aritua, B., Smith, N. J., & Bower, D. (2009). "Construction client multi-projects – A complex adaptive systems perspective." *International Journal of Project Management 27*, 72–79.

Barney, J. B. & Hesterly, W. S. (2012). *Strategic management and competitive advantage: Concepts and cases.* Upper Saddle River, New Jersey: Pearson, Prentice Hall.

Cooper, R. G. & Edgett, S. J. (2003). "Overcoming the crunch in resources for new product development." *Research Technology Management. 46*(3), 48–58.

Cooper, R. G., Edgett, S. J., & Kleinschmidt, E. J. (2001). *Portfolio management for new products.* Cambridge, MA: Perseus.

de Oliveira Lacerda, R. T., Ensslin, L., & Ensslin, S. R. (2011). "A performance measurement framework in portfolio management: A constructivist case." *Management Decision 49*(4), 648–668.

Eisenhardt, K. M. & Martin, J. A. (2000). "Dynamic capabilities: What are they?" *Strategic Management Journal 21*(10/11), 1105–1121.

Elonen, S. & Artto, K. A. (2003). "Problems in managing internal development projects in multi-project environments." *International Journal of Project Management 21*(6), 395–402.

Hansen, L. K., & Svejvig, P. (2018). *Agile project portfolio management, new solutions and new challenges: preliminary findings from a case study of an agile organization. IRIS41/SCIS9 Conference 2018.*

Helfat, C. E., Finkelstein, S., Mitchell, W., Peteraf, M. A., Singh, H., Teece, D. J., & Winter, S. G. (2007). *Dynamic capabilities: Understanding strategic change in organizations.* Malden, MA: Blackwell Publishing.

Jonas, D. (2010). "Empowering project portfolio managers: How management involvement impacts project portfolio management performance." *International Journal of Project Management 28*, 818–831.

Kaufmann, C., Kock, A., & Gemünden, H. G. (2020). "Emerging strategy recognition in agile portfolios." *International Journal of Project Management 37*(7), 429–440.

Kester, L., Griffin, A., Hultink, E. J., & Lauche, K. (2011). "Exploring portfolio decision-making processes." *Journal of Product Innovation Management 28*(5), 641–661.

Killen, C. P. (2017). "Managing portfolio interdependencies: The effects of visual data representations on project portfolio decision making." *International Journal of Managing Projects in Business 10*(4), 856–879.

Killen, C. P. (2022 forthcoming). "Chapter 16: Visualising data for portfolio decision making." In K. Angliss & P. Harpum (Eds.), *Strategic portfolio management in the multi-project and program organisation.* Abingdon, Oxfordshire: Routledge.

Killen, C. P., & Drouin, N. (2017). "Project portfolio management: A dynamic capability and strategic asset for organisational project management." In S. Sankaran, R. Muller, & N. Drouin (Eds.), *Handbook of organizational project management* (pp. 55–69). Cambridge University Press.

Killen, C. P., Geraldi, J., & Kock, A. (2020). The role of decision makers' use of visualizations in project portfolio decision making. *International Journal of Project Management 38*(5), 267–277.

Killen, C. P. & Hunt, R. A. (2010). "Dynamic capability through project portfolio management in service and manufacturing industries." *International Journal of Managing Projects in Business 3*(1), 157–169.

Killen, C. P., Hunt, R. A., & Kleinschmidt, E. J. (2007). *Dynamic capabilities: Innovation project portfolio management. Proceedings of ANZAM 2007*, Sydney, Australia, Australia and New Zealand Academy of Management.

Killen, C. P., Hunt, R. A., & Kleinschmidt, E. J. (2008a). "Project portfolio management for product innovation." *International Journal of Quality and Reliability Management 25*(1), 24–38.

Killen, C. P., Hunt, R. A., & Kleinschmidt, E. J. (2008b). "Learning investments and organisational capabilities: Case studies on the development of project portfolio management capabilities." *International Journal of Managing Projects in Business 1*(3), 334–351.

Killen, C. P. & Kjaer, C. (2012). "Understanding project interdependencies: The role of visual representation, culture and process." *International Journal of Project Management 30*(5), 554–566.

Killen, C. P., Sankaran. S., Clegg, S. and Smyth, H. J. (2022 forthcoming). "Aligning construction projects with strategy." In S. Addyman & H. J. Smyth (Eds.), *Construction project organising.* Chichester: Wiley-Blackwell.

Kopmann, J., Kock, A., Killen, C. P., & Gemünden, H. G. (2017). "The role of project portfolio management in fostering both deliberate and emergent strategy." *International Journal of Project Management 35*(2017), 557–570.

Levine, H. A. (2005). *Project portfolio management: A practical guide to selecting projects, managing portfolios, and maximizing benefits.* San Francisco, CA. Jossey-Bass; John Wiley distributor.

March, J. G. (1991). "Exploration and exploitation in organizational learning." *Organization Science 2*(1), 71–87.

Martinsuo, M. & Lehtonen, P. (2007). "Role of single-project management in achieving portfolio management efficiency." *International Journal of Project Management 25*(1), 56–65.

McDonough III, E. F. & Spital, F. C. (2003). "Managing project portfolios." *Research Technology Management 46*(3), 40–46.

Mikkola, J. H. (2001). "Portfolio management of R&D projects: Implications for innovation management." *Technovation 21*(7), 423–435.

Petit, Y. (2012). "Project portfolios in dynamic environments: Organizing for uncertainty." *International Journal of Project Management 30*(5), 539–553.

Petro, Y. (2017). *Ambidexterity through project portfolio management.* Newtown Square, PA: Project Management Institute

Rad, P. F. & Levin, G. (2008). "What is project portfolio management?" *AACE International Transactions*: TC31.

Saeed, M. A., Jiao, Y., Zahid, M. M., Tabassum, H., & Nauman, S. (2021). "Organizational flexibility and project portfolio performance: The roles of innovation, absorptive capacity and environmental dynamism." *International Journal of Managing Projects in Business 14*(3), 600–624.

Sicotte, H., Drouin, N., & Delerue, H. (2014). Innovation portfolio management as a subset of dynamic capabilities: Measurement and impact on innovative performance. *Project Management Journal*, 45(6), 58–72.

Smith, K. A., Vasudevan, S. P., & Tanniru, M. R. (1996). "Organizational learning and resource-based theory: An integrative model." *Journal of Organizational Change Management 9*(6), 41–53.

Stettina, C. J., & Hörz, J. (2015). Agile portfolio management: An empirical perspective on the practice in use. *International Journal of Project Management 33*(1), 140–152.

Teece, D. J. (2007). "Explicating dynamic capabilities: The nature and microfoundations of (sustainable) enterprise performance." *Strategic Management Journal 28*(13), 1319–1350.

Teece, D. J., Pisano, G., & Shuen, A. (1997). "Dynamic capabilities and strategic management." *Strategic Management Journal 18*(7), 509–533.

Tushman, M. & O'Reilly, C. (1996). "Ambidextrous organizations: Managing evolutionary and revolutionary change." *California Management Review 38*(4), 8–30.

Tushman, M., Smith, W., Wood, R., Westerman, G., & O'Reilly, C. (2002). *Innovation streams and ambidextrous organizational designs: On building dynamic capabilities.* Organisational Studies Group Seminar, Cambridge, MA: Massachusetts Institute of Technology.

Van der Hoorn, B. (2021). *Visuals for influence: In project management and beyond.* Toowoomba: USQ Press. https://usq.pressbooks.pub/visualsforprojectmanagement/

Wang, C. L. & Ahmed, P. K. (2007). "Dynamic capabilities: A review and research agenda." *International Journal of Management Reviews 9*(1), 31–51.

Warglien, M. (2010). *Seeing, thinking and deciding: Some research questions on strategy and vision. Workshop on the Power of Representations: From Visualization, Maps and Categories to Dynamic Tools, Academy of Management Meeting,* August 6th, 2010, Montreal.

Index

Pages in *italics* refer figures, pages in **bold** refer tables, and pages followed by n refer notes.

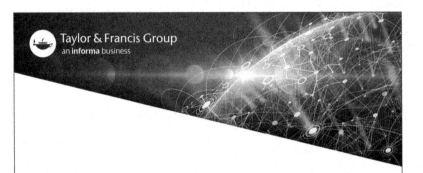

Printed in the United States
by Baker & Taylor Publisher Services